Practical Approach of Quality Assurance in Spinning Mills

Practical Approach of Quality Assurance in Spinning Mills

S. N. Mukherjee
P. K. Majumdar

WOODHEAD PUBLISHING INDIA PVT LTD

New Delhi

Published by Woodhead Publishing India Pvt. Ltd.
Woodhead Publishing India Pvt. Ltd.,
303, Vardaan House, 7/28, Ansari Road,
Daryaganj, New Delhi - 110002, India
www.woodheadpublishingindia.com

First published 2020, Woodhead Publishing India Pvt. Ltd.
© Woodhead Publishing India Pvt. Ltd., 2020

Woodhead Publishing India Pvt. Ltd. ISBN: 978-93-88320-29-0
Woodhead Publishing India Pvt. Ltd. e-ISBN: 978-93-88320-30-6

Typeset by Bhumi Graphics, New Delhi
Printed and bound by Replika Press Pvt. Ltd.

Contents

Preface

The final goal of any production and manufacturing process is to produce reliable products of desired quality, as a finished result. If the quality of the product does not comply with the expectations and the needs of the consumers and does not meet international standards, then the production might as well have been a waste of time and resources. . Although this sounds harsh, it is indeed important to make sure as a manufacturer to provide a high-quality product meeting the requirements of the customer. This is relevant and applicable in every sphere of the economy and the Industry of Textile production just cannot live without it. Textile is one of the major sectors of Indian Industries, the scope of which is increasing day by day. Textile has grown up to the peak of the consumer needs which requires the assuredly, quality & sustainable characteristics to further meet the competency in the market. In today's highly competitive and changing consumer market, those in the soft-lines and textile industry, including manufacturers, brands and retailers, need to be sure that the products they deliver to market meet consumer expectations along with the required safety, regulatory and government standards globally.

The quality of a product or process should be checked before it is put into large-scale usage. The quality of the product, its performance, and its reliability are the key factors while testing is performed. Testing can be defined as the methods or protocols adopted to verify/determine the properties of a product. It can be divided primarily into two types: regular process testing and quality assurance testing. Routine testing helps to streamline the daily process. Quality assurance testing helps the process or product in the long run to establish credibility. Testing can also be defined as the procedures adopted to determine a product's suitability and quality.

Inspection is the source of data which, analyzed and interpreted through statistical methods, is continuously fed back to production people for corrective and preventive action. Inspection is thus the act of screening out defectives before they reach the customer.

The need for a book exclusively on Quality Assurance in spinning has been conceived by us for a long time. While a good number of books are available on Quality Control and Quality Management, a book exclusively dealing with Quality Assurance in Spinning is still not available. In this book, attempts have been made by the authors to describe importance of quality assurance in spinning in present day's context. Important areas related to quality assurance viz. mixing, inspection & checking at different stages of spinning, testing & interpretation of test results, application of different basic statistical tools, identification and remedy of different yarn and fabric defects etc. have been covered.

This book is the outcome of the long time desire of the authors to combine their long industrial and teaching/research experience and share with others in industry and academic field. This desire will only be fulfilled if the readers get benefit out of it. This book is written not only for quality assurance people but production people can also fetch advantage of it.

Lastly, we must acknowledge the inspiration and encouragement we received from our wives Mrs. Ratna Mukherjee and Mrs. Sunanda Majumdar while writing this book. We must also acknowledge the encouragement and support received from our daughters Dr. Tinni Chaudhuri & Nabanita Chatterjee(Daughters of S. N. Mukherjee) and Dr. Sanghamitra Majumdar(Daughter of P. K. Majumdar)

S. N. Mukherjee

P. K. Majumdar

Introduction

Present day's philosophy in manufacturing sectors in India is – "Think globally and produce globally". Total dependence on domestic market hardly brings any prosperity for any growing manufacturing sector in the long run. Textile industry is not an exception to it. It is thus very much important to know how to produce globally – the answer is "Quality", which needs to be assured. Quality product manufacturing involves a perfect team work with standardisation in raw material and process norms. Use of good machinery, trained labour force and lastly a very strong team of quality control are essential requirements for manufacturing quality products. Although statistical quality control (SQC) is used as a tool to prevent faulty products reaching the customer through inspection of samples of work-in-progress and finished goods to ensure standards are being met, it is not enough to assure quality. Also if defect levels are very high, the company's profitability will suffer unless steps are taken to tackle the root causes of the failures. The activity of the SQC department in most cases is merely related to fault finding due to which it develops an uneasy relationship with the people in production department as well as with the labour forces too. Also, a major problem is that individuals are not necessarily encouraged to take responsibility for the quality of their own work. However, this old system of SQC's working is waning out very fast. Now, many mills (here only spinning mills are discussed) have adopted a most effective system of quality checking through "Quality Assurance Department" which assures that quality of the ultimate products are within standard norms by following a planned, systematic and proven system of manufacturing. Quality assurance systems emphasise catching defects before they get into the final product.

The authors, in this book has discussed in details on how the quality assurance department should work in spinning mills.

1
Quality assurance in spinning mills

1.1 What is quality assurance?

Quality assurance (QA) is a way of preventing mistakes or defects in manufactured products and avoiding problems when delivering solutions or services to customers; which is defined by ISO 9000[1] as "part of quality management focused on providing confidence that quality requirements will be fulfilled".

Quality is meeting the requirement, expectation and needs of the customer being free from defects, lacks and substantial variants. There are standards needs to follow to satisfy the customer requirements.

Assurance is provided by organisation management, it means giving a positive declaration on a product which obtains confidence for the outcome. It gives a security that the product will work without any glitches as per the expectations or requests.

The terms "quality assurance" and "quality control" are often used interchangeably to refer to ways of ensuring the quality of a service or product [2]. Quality assurance comprises administrative and procedural activities implemented in a quality system so that requirements and goals for a product, service or activity will be fulfilled [2]. It is the systematic measurement, comparison with a standard, monitoring of processes and an associated feedback loop that confers error prevention [3]. This can be contrasted with quality control, which is focused on process output. Quality assurance includes two principles: "Fit for purpose" (the product should be suitable for the intended purpose) and "right first time" (mistakes should be eliminated). QA includes management of the quality of raw materials, assemblies, products and components, services related to production and management, production and inspection processes [4].

1.2 How to achieve the goal?

High levels of quality are essential to achieve business objectives. Quality, a source of competitive advantage, should remain a hallmark of any company's products. High quality is not an added value; it is an essential basic requirement. Quality does not only relate solely to the end products but also relates the work processes and the way they are being executed. The work processes should be as efficient as possible and continually improving. Employees of any organisation constitute the most important resource for improving quality. Each employee in all organisational units is responsible for ensuring that their work processes are efficient and continually improving.

Top management should provide the training and an appropriate motivating environment to foster teamwork within organisational units for employees to improve processes. Ultimately, everyone in an organisation is responsible for the quality of its products.

1.2.1   Practical approach to reach the goal

At the very first step, a perfect work culture is to be developed among all staff of QA department. The head of the QA department is fully responsible for developing and creating this work culture. The head should get him more involved in day to day work than the subordinates and should be in touch with each and every subordinate staff in workplace looking and asking their observations of studies and their opinions. This creates a sense of responsibility among subordinates. Also a format is to be prepared and to be distributed among all staff members to fill-up. This is only one time exercise. A sample format is given here.

Format

1. Name .... 2. Age ..... 3. Qualification ...... 4. Total Experience ....

Daily routine jobs

S. No.	Nature of studies	Time required to execute a study	Total time
1.	Full doff breakage of a simplex m/c	1 h 40 min	1 h 40 min
2.	Simplex m/c's wrapping (Total 8 machines)	1 h	2 h 40 min
3.	Drawing m/c's wrapping	1 h	3 h 40 min
4.	----------------------------	-------------	---------------
5.	----------------------------	-------------	---------------
		Total	**7 h**

The type of study and time needed to complete the study is to be given by the QA head against each staff. Out of 8 h duty, 30 min is to be given to the each staff (call staff as investigator) for personal requirements and 30 min is to be given for writing the shift's report. Printed format for each and every study is to be maintained daily. Two copies of each study are to be made daily viz. One copy for production head and another copy for maintenance head. Daily a meeting among the heads of production, maintenance and quality assurance departments to be carried out in the evening to discuss about the quality assurance study report for action to be taken by the heads of the three departments. Once this format is created and adhered to, then the accountability and responsibility of the investigators are going to increase certainly. All the investigators should be instructed as follows:

1. Report immediately to production persons and QA head when any odd observation is found during study. Never wait for the completion of the study.

2. Importance of each study and its subsequent impact on forward process should be made clear to them and if any help is needed from head, they should not hesitate to take that.

3. Always try to find out the root cause of any problem and give a suggestion for overcoming. Do not act as a fault finder.

1.2.2   Role of QA head

A proper atmosphere of competitiveness among all investigators should be created by the head of the department and he should act as their mentor or guardian. QA head should remember that for any success of the case study or any job the person concerned should be given proper credit and for any failure of any case study or any job, responsibility should be borne by him. Each person in QA department should know that the work is more important than personal difference.

1.2.3   Motivation

Motivation in day to day job is very important. It can be done in two ways, viz. (1) through monetary, cash prizes, promotion, etc. and (2) by recognising the individual efforts. The first part lies with the top management's decision with financial involvements. The second option lies on the head of QA department. "Good work" appreciation should be given to the person concerned, in the presence of all staff of QA department. Worth reading technical literatures from different textile magazines or research papers published in technological conferences are to be highlighted among all staff by QA head in an abridged form to increase the technical aptitude of the staff. One important slogan which is very appropriate should be displayed in QA department and everybody should follow it. Follow 3G (a Japanese slogan):

1. Genchi – Go to the spot.
2. Genbutsu – See the actual problem.
3. Genjitsu – Take realistic action based on the fact.

1.2.4   Checking

Lastly, a proper checking system for routine jobs is to be developed at different stages of production to plug any loop hole which generate sub-standard quality. This quality checking system should be based on 3 tiers checking system as follows.

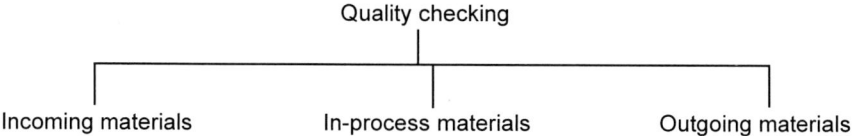

- "Incoming materials" consist of either cotton or synthetic/man-made fibres.
- Checking of "In-process materials" consist of monitoring the process parameters at each stage of production to see that it is adhered to the standard norm up to the final cone winding stage.
- Checking of "Outgoing materials" is similar to in-process but in the packing stage.
- The frequency of checking at different stages may be based on daily, weekly and monthly basis as per the requirements which will be discussed later on.

1.2.5   Maintenance

Close observations on maintenance schedule changes/jobs which have a direct impact on quality, are to be monitored. This monitoring, which are mostly paper works, is to be carried out by the head of QA department.

The important schedule changes/jobs are:

Carding:
1. Sharpening of cylinder wires and flat wires and grinding of doffer wires.
2. Replacement of Licker-in wires.
3. Replacement of cylinder and doffer wires.
4. Replacement of flat tops.
5. Replacement of combing segments and stationary flats (front and back).

Drawing:

1. Top cots buffing.
2. Replacement of top cots after reaching a minimum permissible diameter.
3. Checking of nip load pressure.

Simplex:

1. Top cots buffing.
2. Top cots replacement.
3. Checking of top arm pressure (front and back).
4. Top and bottom aprons change.
5. Twist master change.

Ring frame:

1. Top cots buffing
2. Top and bottom aprons change.
3. Top front cots change at minimum permissible diameter.
4. Diameter difference between front and back cots.
5. Checking of top arm pressure checking (front top only).

Recommendation of grinding schedules for wires and its replacement for high production cards (for cotton only) are given here.

Taker-in wire replacement – After 150 tons of production.

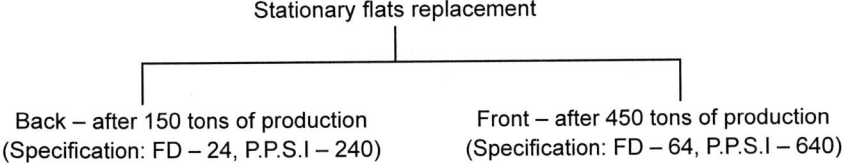

Stationary flats replacement

Back – after 150 tons of production (Specification: FD – 24, P.P.S.I – 240)   Front – after 450 tons of production (Specification: FD – 64, P.P.S.I – 640)

Combing segments (Licker-in) replacement – After 300 tons of production.
Cylinder wires replacement – After 450 tons of production.
Doffer wires replacement – After 450 tons of production.
Flats wires replacement – After 450 tons of production.

Sharpening/grinding schedule:

Cylinder   :   1st after 150 tons of production.
                         2nd after 300 tons of production.
                         3rd after 400 tons of production.

Flats	:	1st after 150 tons of production.
		2nd after 300 tons of production.
		3rd after 400 tons of production.
Doffer	:	1st after 230 tons of production.
		Last after 400 tons of production.

The above change schedule, buffing and grinding schedule, etc. are to be taken from the maintenance department in printed form. This procedure is mostly followed by the maintenance department. The head of QA department should keep a track on it to observe the proper follow-up by the maintenance department. In this follow-up process, certain special attention is to be put on the particular machine where any change as per schedule is due. For example, the cylinder wires change schedule of a particular card has fallen due as per schedule. Here QA head should convey the production department to make yarn sample from the particular card vs. any card in which schedule change has not come yet. Then three yarn parameters like neps, thick and thin places per 1 km of the two sample yarns prepared using the two cards are to be compared. Change of wires are advisable only when any deterioration is observed in the said parameters for yarns prepared using the card whose wire change is due. In fact for evaluating carding action, the difference between thick and thin places is very important which reflects actual number of thick place created by carding.

This process is to be repeated for 2–3 times to check whether the same trend is coming or not. While making comparative yarn samples, it is to be remembered that for any such comparison, yarns processing from card up to ring frame are to be carried out on same machine, same delivery (for drawing), same spindle (for simplex and ring frame) to avoid any machine variance. If the values of the tested parameters of the yarn samples for both the cards are same, then no change of the wires is required. This process is to be repeated from time to time till differences in values of the parameters are observed.

Likewise, for twist master change, 60 yd roving lea strength is checked and compared with the same from any other simplex m/c (where change is not due) running with same roving hank/tpi/spindle speed/rachet wheel/winding wheel, etc. If 60 yd lea strength of the "schedule change simplex" is found lower than the other one, then twist master change is to be made, otherwise no change is required.

For cots buffing in simplex and ring frame, yarn $U\%$ and imperfections are to be checked in the same manner as stated above. Same procedure to be applied for aprons change. QA department should always take a techno-economic attitude and try to save time and money where ever possible. In

India ROI (Return on Investment) in textile industries is low with respect to other engineering industries. On an average, ROI comes as 5–6% and therefore, saving attitude is extremely important at all stages of productions and maximum efforts should be given to reduce hard wastes and unusable wastes.

1.3   Cotton fibre

Cotton has been a major agricultural crop and a significant fibre in the world textile market. Cotton is a natural plant fibre which grows around the seed of the cotton plant. Fibres are used in the textile industry, where they are the starting point of the production chain. First, the cotton fibre is obtained from the cotton plant and then spun into yarn. From there, the cotton yarn is woven or knitted into fabric. The use of cotton has a long tradition in the clothing industry due to its desirable characteristics. Cloths made of this fibre are moisture-absorbent, have a good drape and are known for their long durability. It is generally recognised that most consumers prefer cotton personal care items to those containing synthetic fibres.

1.3.1   History of cotton cultivation

The earliest evidence of using cotton is from India and the date assigned to this fabric is 3000 B.C. There were also excavations of cotton fabrics of comparable age in Southern America [5].The trail of cotton's commercial growth began in 1793 with Eli Whitney's invention of the cotton gin. The gin was designed to mechanically separate cotton fibres, or lint, from cottonseeds. The gin increased the market value of cotton in the 10-year period from 1793 to 1803 from $150,000 to more than $8 million [6]. This proved to be just the beginning of the mechanisation of an agricultural crop, from planting and harvesting to processing, spinning and product manufacture. The first patent for a mechanical cotton picker was filed in 1850 [7].The cotton gin and picker joined the development of mechanical yarn-spinning equipment during the Industrial Revolution in the United Kingdom and the United States during the early 19th century [8]. By the mid-20th century, mechanical harvesting and ginning came into widespread use in the cotton-growing areas of the United States. These technologies were the forerunners of the modern, high-speed harvesting, ginning, yarn processing and spinning technologies used in today's cotton industry.

From the 19th century, when cotton was grown and consumed within a few countries or geographical areas [6] to the 21st century when it is grown and traded from more than 100 countries, cotton has increased in production and significance in international trade.

World textile fibre consumption in 1998 was approximately 45 million tons. Of this total, cotton represented approximately 20 million tons [9]. Total fibre consumption has increased every year except the global recession year of 2008, growing from 37.9 million tons in 1990 to 89.7 million tons in 2015, an average increase of 2.1 million tons per year. Over this same period, cotton consumption has increased from 18.6 million tons to 24.7 million tons, for an average increase of 244 thousand tons per year. However, cotton consumption peaked at 26.6 million tons in 2007 and has since exhibited an average decrease of 238 thousand tons per year. Between 1990 and 2013, cotton's share of the global fibre market decreased from about 50% to about 28% [10]. During 2016 and 2017 global cotton consumption was 24.52 and 25.22 million tons, respectively.

Cotton cultivation first spread from India to Egypt, China and the South Pacific. Even though cotton fibre had been known already in Southern America, the large-scale cotton cultivation in Northern America began in the 16th century with the arrival of colonists to southern parts of today's United States [5].

Cotton has been a major agricultural crop and has become a significant fibre in the current world textile market where, between 2000 and 2004, an average of 20.1 million metric tons was produced by cotton-growing countries [11]. World cotton production was 18.96 million tons in 1990 and maintained steady growth over the years except drop in production in 2008 and 2009 to 23.60 and 22.50, respectively [12].The world cotton market experienced dramatic developments in the first half of the 2015 marketing year caused by an acute drop in production – about 9% – in major producing countries. Worldwide cotton production has not declined this much since 2008. Production fell in almost all major cotton producing countries led by Pakistan, the United States and China, which experienced declines of 5%, 19% and 17%, respectively. Adverse weather, lower global world market demand and policy uncertainty all contributed to the sharp decline. The decreased synthetic fibre prices driven by substantially lower oil prices placed huge competitive pressures on world cotton markets [13]. After the sharp drop in production in 2015/16, the 2016/17 production recovered by 7% to 23 million tons. International Cotton Advisory Committee (ICAC) indicated that recovery continues in the period 2017/18, with production indicated at 25.4million tons due to increased area [14].

1.3.2   Varieties and types of cotton

There are many different varieties and types of cottons. Their characteristics determine the use for the cotton, and hence its value. Cotton is a member of the order Malvales, family Malvaceae, genus *Gossypium*. The genus

Gossypium consists of 50 wild and cultivated species, out of which only four are grown on a commercial scale in the world. *Gossypium hirsutum*and *Gossypium barbadense* are called New World species and account for about 95% and 3% of world production, respectively. *Gossypium arboreum* and *Gossypium herbaceum* are called Old World or Asiatic cottons and are grown commercially in India, Pakistan and parts of South-east Asia, accounting for about 2% of world production.

Extra-long staple Egyptian, American Egyptian or Pima and Sea Island cotton belong to the species *G. barbadense*. The fibre in this group is long, fine and strong with a staple length in excess of 32 mm (1–1/4″), a micronaire value below 4.0 and strength of up to 40 g/tex. The fibre of Old World cottons is generally shorter than 25 mm (1″) and coarse, with a micronaire value above 6.0. Worldwide about 500 varieties are used for commercial cotton production. Most of them are Upland species.

According to International Cotton Advisory Committee (ICAC), world cotton supply can be divided into six categories based on commonly perceived competitive relationships between cottons of differing quality, variety and geographic origins: extra-fine, fine, high-medium, medium, coarse count and waste/padding. The categories are roughly parallel to staple length categories but are designed to incorporate more than just staple length information because two cottons of equal length might actually have significantly different spinning characteristics [15].

1.3.3   Major cotton producing countries

India, China, United States and Pakistan –account for about over 70% of world cotton production, with India and China each accounting for about one-fourth of world cotton production. Out of the world's leading cotton producing countries India amounted to around 6,205 metric tons and was the largest producer of cotton worldwide in crop year 2017/18 [16].

The following statistic shows cotton production by the world's leading cotton producing countries in crop year 2017/18 [16].

Production in thousand metric tons:

India	6,205
China	5,987
United States	4,555
Brazil	1,894
Pakistan	1,785
Australia	1,045

Turkey	871
Uzbekistan	838
Turkmenistan	296
Burkina	158

1.3.4   Brief ideas on the cultivation of cotton in India

India is the country to grow all four species of cultivated cotton *G.arboreum* and *G.herbaceum* (Asian cotton), *G.barbadense* (Egyptian cotton) and *G. hirsutum* (American Upland cotton). *Gossypium hirsutum* represents 88% of the hybrid cotton production in India and all the current *Bt*-cotton hybrids are *G.hirsutuim.*In India, majority of the cotton production comes from nine major cotton growing states, which are grouped into three diverse agro-ecological zones, Northern zone comprising States of Punjab, Haryana and Rajasthan, Central zone comprising the States of Gujarat, Maharashtra and Madhya Pradesh and Southern zone comprising the States of Telangana, Andhra Pradesh and Karnataka. Besides this, cotton is also grown in the States of Tamil Nadu and Orissa [17].

State wise cotton production in 2018 [18]:

State Production in lakh bales (1bale weighs 170 kg)

Gujarat	125
Maharashtra	85
Telangana	50
Andhra Pradesh	27
Karnataka	28
Haryana	25
Madhya Pradesh	18
Rajasthan	17
Punjab	14
Tamil Nadu	5
Orissa	3

1.3.4.1   Different cotton station in India

Maharashtra – Wani, Warora, Panbarkawda, Ghatani, Paratwada, Jalgaon, Dhule, Akola, Khamgaon, Nagpur, Nanded, Parbhani, Aurangabad, Phalton and Baramati.

Types of cotton produced are Mech-1, H-4, LRA, Jkay-1, Y-1, MCU-5, DCH-32, etc.

Gujrat – Porbandar, Dhoraji, Manvadar, Surinder Nagar, Limbdi, Anjar, Palej, Jahangerpura, Kharjam Karavan, Miyagaon, Bodeli, Kaledia, Himatnagar, Botad and Dholka.

Types of cotton produced are S-6, Kalyan, Wagad, CO-2.

MP – Barhanpur, Khandua, Bhikengaon, Sandhwa, Dhamnod, Badnavar, Ratlam, Khargaon and Sansar.

Types of Cotton produced are MCU-5, DCH-32, Mech-I, LRA.

Punjab – Fazilka, Muktsar, Jaitu, Abohar, Kotkapura, Bhatinda, Bhuchu, Budhalda and Rampuraphool.

Types of cotton produced are mainly J-34, S/G and R/G.

Haryana – Sirsa and Fathehabad.

Types of cotton produced are mainly J-34 and R/G.

Rajasthan – Sri Ganganagar, Kesrisingpur, Gajsinghpur and Anupgarh.
Types of cotton produced are mainly Desi andJ-34.

1.3.4.2   Common varieties of cotton available from different stations usually found in cotton mills in India

1. DCH-32: Haveri, Hubli, Ranebennur, Dhanwad, Narendra, Sendhwa.
2. MCU-5: Sansar, Jalna, Arvi, Warud, Kothaguda, Devalgaon, Malkapur, Errupalem, Sillud, Guntur, Sendhawa.
3. H-4: Pandhurana, Dhamnod, Sendhawa, Sansar.
4. S-6: Kadi, Wankaner, Vijapur, Murbi, Botad, Gudawari, Himatnagar, Tankara, Gundal, Harij, Bhuj, Dhasa, Shapok, Radhanpur, Chaila, Anjar, Talaja, Gariyadhar.
5. BB (Bunny Brambha): Burgampadu, Mylavaram, Warangal, Nandigama, Enkur, Prathipadu, Phirangipu, Madhira, Sattenapal, Chityal, Marangal, Palvancha.
6. Mech-1: Pandhurana, Sansar, Wradha, Burtianpur, Khargaon.

1.3.4.3   Properties of few popular Indian cotton are given in Table 1.1

Table 1.1: Properties of few popular Indian cotton

Varieties	Trash%	S.F %	Rd	+b	P.M%	UHML (mm)	ML (mm)	UI	Bundle St. (g/tex)	Mic.
S-6(R/G)	3.90	5.20	77.5	8.8	80	30.8	26.2	84	31.50	4.50
J-34	5.50	2.10	70	8.5	86	30	25.8	86	31.80	5.20
Mech-1 R/G	3.50	3.40	73	9.0	82	31.2	26.2	84	32.0	4.50
MCU-5	2.50	1.80	73	9.20	82	33.0	28.5	86	33.0	3.80
DCH-32	1.80	2.0	71	9.80	70	34.5	29.1	85	37.5	3.30
PIMA (Imported)	1.2	(−)3.1	70.5	11.2	84	36.3	31.7	88	42.8	4.20
GIZA-86	1.4	(−)2.8	82.0	8.7	88	34.8	30.6	88	40.5	4.0

1.4   References

1. ISO 9000:2005, Clause 3.2.11ISO 9000:2005,

2. Quality Assurance vs Quality Control – Learning Resources – ASQ.

3. The Marketing Accountability Standards Board (MASB), Common Language in Marketing Project.

4. Stebbing, L., *Quality Assurance: The Route to Efficiency and Competitiveness*, 3rd edition, Prentice Hall, 1993,p. 300. ISBN978-0-13-334559-9.

5. Tortora, P.G., and Collier, B.J., *Understanding Textiles*, 5th edition, Prentice-Hall, 1997.

6. *Joseph's Introductory Textile Science*, Sixth Edition, Ed. P. Hudson, A. Clapp, and D. Kness, Harcout Brace Jovanovich College Publishers, 1993, p. 53.

7. *American Cotton Handbook*, Ed. D. Hamby, Enterprise Publishers, Third Edition, Vol. 1,1965, p. 53.

8. Yafa, S., *Big Cotton*, S. Penguin Group, 2005, pp. 70–146.

9. Shaw, L.H., *Cotton's Importance in the Textile Industry*, Symposium, Lima, Peru, May 12, 1998.

10. Ethridge, D., *Policy-Driven Causes for Cotton's Decreasing Market Share of Fibres*,33rd International Cotton Conference, Bremen, March 16–18,2016, pp. 1–28.

11. Cotton Incorporated, Monthly Economic Letter, June 10, 2005.

12. Cotton Production Worldwide 2018, Statistic,https://www.statista.com/statistics/259392/cotton-production-worldwide-since-1990/.

13. OECD-FAO Agricultural Outlook 2016–2025, http://dx.doi.org/10.1787/agr_outlook-2016-en.

14. Wright, B., World Cotton Production Seenup11% in2017/18, 2018. https://www.just-style.com/news/world-cotton-production-seen-up-11-in-201718_id132520.aspx.

15. Types of Cotton, Cotton Guide,http://www.cottonguide.org/cotton-guide/market-segments-types-of-cotton/.

16. Cotton Production by Country Worldwide, 2018, https://www.statista.com/statistics/263055/cotton-production-worldwide-by-top-countries/.

17. Cotton Sector-Ministry of Textiles,texmin.nic.in/sites/default/files/Textiles_Sector_Cotton.pdf.

18. Zinn, R., Top 10 Largest Cotton Producing States in India,www.trendingtopmost.com/worlds.../ largest-cotton-producing-states-india-world/.

2

Mixing

Mixing is defined as the intermingling of different grades of same fibres together to achieve desired quality of the end product at the most economic cost. There is no particular ratio of fibres to mix with each other. In cotton fibre spinning, raw cotton is the prime factor that influences the quality of yarn. The main technological challenge in any textile process is to convert the high variability in the characteristics of input fibres to a uniform end product. This critical task is mainly achieved in the mixing process. Mixing department in the spinning mill plays a crucial role in the formulation of appropriate mix of fibres. Mixing has a significant impact on the end-product cost and quality. Conventionally, mixing could be thought of combining of fibres together in somewhat haphazard proportions whose physical properties are only partially known so that the resultant mixture has only generally known average physical properties which are not easily reproducible.

In the modern context of spinning mixing is performed by instruments which are programmed to provide particular information on fibre for scientific selection of raw material. This selection process meets the quality requirement set by the customer.

2.1   Selection of cotton

Today, in the times of globalisation it is required to produce yarn and fabrics of required quality and economic prices. This is possible with complete integration of fibre quality in the process of yarn manufacturing. This is related to the process of cotton fibre selection and cotton fibre blending. The quality of final yarn is largely influenced (80%) by the characteristics of raw cotton [1]. Prior to the development of fibre testing equipment, three fibre parameters have been used to determine the quality value of cotton fibre. These are grade, fibre length and fibre fineness. The expertise and experience played a dominant role, making the formulation of mix, highly subjective. One of the common approaches was massive blending, in which vast qualities of bales were mixed by grade or growth area to reduce variability [2]. The development of fibre testing instruments such as the high volume instrument (HVI) and the advanced fibre information system (AFIS) has revolutionised the concept of fibre testing. With the HVI it is pragmatically possible to determine most of the quality characteristics of a cotton bale within 2 min.

Based on the HVI results, composite indexes such as the fibre quality index (FQI) and spinning consistency index (SCI) can be used to determine the technological value of cotton; this can play a pivotal role in an engineered fibre selection programme [3, 4].

The mathematical linear programming was used for cotton cost optimisation. Though this approach was fundamentally sound, it was not accessible easily because of slow fibre testing and lack of powerful computing system for solving the prolonged statistics [5, 6]. Linear programming method earlier used the simplex method which has been mathematically stabilised and permits sensitivity and parametric analysis. Today use of computer and software programs has made mathematical complexity an easy task [7–10]. LP models take various fibre mixing quality constraints into account in determining the best selection and mixing of cottons. However, they are not always particularly effective, because their restrictive structures lead to over-simplistic representations of the real system. The optimal solution provided by a LP model is therefore not necessarily accurate in practice.

The decisions involved in cotton mixing are based on multiple requirements or goals (such as optimal strength, optimal cost and so on), and the management may choose to relax one requirement to examine the impact that it has on a different requirement; for example, the strength requirement may be relaxed by 10% to examine the cost implications. Goal programming (GP), which allows the inclusion of goals and sub-goals in its calculations in order to achieve a satisfactory level for the individual user, has therefore proven beneficial in the textile industry [11].

- The GP approach offers the following advantages over LP:
- Deviations from the target value are minimised.
- Goals are ranked based on their contributions.

The desired level of satisfaction is achieved based on the ranking assigned to the different goals and the minimisation of deviations from these goals.

Cotton is a natural fibre having variability galore in its properties. Most of these properties play a decisive role in determining the tensile and evenness characteristics of spun yarns. For example, yarn strength, which is considered to be the most important property of spun yarns, is largely influenced by the tenacity, elongation, length, length uniformity, short fibre content and fineness (micronaire or microgram/inch) of constituent cotton. For ring spun yarns, the contribution of fibre tenacity, elongation, length, length uniformity and micronaire to yarn tenacity is 20%, 5%, 22%, 20% and 15%, respectively [1]. Therefore, the selection of suitable cotton fibres gives rise to a situation, which involves multi criteria decision making (MCDM).

In a simple MCDM situation, all the criteria are expressed in terms of same unit (e.g., kg or dollars). However, in many real life MCDM problems, different criteria may be expressed in different units. Examples of such dimensions encompass g, cm, rupees, etc. It is this issue of multiple dimensions, which makes the typical MCDM problem to be a complex one and the analytic hierarchy process (AHP) or its variants, may offer great assistance in solving this type of problem [12–15]. Generally, the overall quality of cotton fibre is measured by a dimensionless parameter known as fibre quality index (FQI). To obtain the FQI value of a cotton fibre the product of 50% span length, bundle strength and maturity is divided by the fineness (micronaire) value [16]. Therefore, in FQI length, strength and maturity of cotton fibre receives equal importance irrespective of their influence on the final yarn quality. Moreover, the contribution of various fibre parameters to yarn quality varies with the type of yarn spinning technology. For example, the influence of fibre length to yarn tenacity is more predominant in ring spun yarns as compared to rotor spun yarns [17, 18]. In a stark contrast to FQI system, the decision maker in AHP could assign different weights to the fibre parameters depending on their importance and separate models could be developed for ring and rotor spun yarns.

The success of the GP model is dependent on the accuracy of the input data, particularly the goal priorities and their weightings. The goal priorities often fluctuate; it may be desirable to relax the target values of some (less important) goals to achieve substantial improvements in another goal.

The other important input is the number of bales in stock and their fibre characteristics, which vary from bale to bale. In most computer based models designed for cotton mixing, average fibre characteristics, based on a few bales, are used, rather than those of all the bales actually used in mixing.

This means that the desired fibre characteristics cannot be guaranteed; however, the alternative, which would be to keep a large inventory of bales that are known to have the required characteristics, is extremely expensive. The solution adopted is the use of a cotton inventory (CI) model, which provides information on the availability of bales and their fibre characteristics during the decision-making process. The qualities available are then used as constraints in the GP model. The average fibre characteristics of the bales actually required in the mix are used in the GP model calculations, which ensure that the actual quality of fibre mixing is close to the desired level. In conclusion, then, an integrated use of GP, AHP(analytic hierarchy process) and CI provides a rational and effective solution to the problems involved in cotton mixing decisions, and helps to ensure the quality of the raw material coming into the blow room [11].

2.1.1   Models to determine the technological values of cotton

Traditionally grade, fibre length and fibre fineness have been used to determine the quality value of cotton fibre. Traditional fibre testing was a slow and tedious, requiring 4–6 h for testing a sample for the various properties. Because of this, only a small proportion of the bales in the lot were tested. In early eighties the introduction of evaluation of fibre properties using testing equipment like Baer sorter, Micronaire, Trash Analyser, etc., made the process less subjective. With the advent of HVI and AFIS the fibre parameters are measured within few minutes and thus, testing of 100% bales puts fibre selection on a scientific basis, rather than a matter of chance. Different models have been developed to determine technological value of cotton, as discussed below:

2.1.1.1   *Fibre quality index (FQI)*

This is probably the most widely used method to determine the technological value of cotton [16, 19–21]. The main reason for its popularity may be attributed to the simplicity of the equation used. Several variants of the FQI model are available. FQI model proposed by the South Indian Textile Research Association [21] is as follows:

$$FQI_{HVI} = \frac{UHML.UI.FS}{FF}$$

where, L is the 2.5% span length, UR is the uniformity ratio, FS is the fibre bundle tenacity, M is the maturity coefficient and FF is the fibre fineness (micronaire). If the HVI mode of fibre testing is used, then the above expression is changed as follows:

$$FQI = \frac{L.UR.FS.M}{FF}$$

where, FQI_{HVI} is the HVI quality index, UHML is the upper half mean length and UI is the uniformity index.

2.1.1.2   *Spinning consistency index (SCI)*

This is a calculation for predicting the overall quality and spinnability of the cotton fibre. It is chiefly used to gain within and between lay-down consistencies of major cotton properties. The regression equation of SCI uses most of the individual HVI measurements, and it is based on the five-year crop average of United States Upland and Pima cotton. The regression equation [22] used to calculate SCI is as follows:

SCI = −414.67 + 2.9FS + 49.17UHML + 4.74UI − 9.32FF+ 0.65Rd + 0.36 (+b)

where, Rd is the reflectance degree and +b is the yellowness of cotton fibre.

2.1.1.3   Premium-Discount Index (PDI)

This method was proposed by Mogazhy et al. [23]. It includes the development of a multiple regression equation relating fibre properties and yarn strength, the determination of the percentage contribution of fibre properties to yarn strength, the selection of a reference set of cotton properties, the determination of a difference factor between the fibre property and the reference set, and finally the development of a premium-discount formula. The regression equation of the following form is developed from the available fibre and yarn data.

$$\textbf{Yarn tenacity} = C_1 + C_2 \cdot \textbf{FS} + C_3 \cdot \textbf{FE} + C_4 \cdot \textbf{UHML} + C_5 \cdot \textbf{UI} + C_6 \cdot \textbf{SFC} + C_7 \cdot \textbf{FF}$$

where, C_1, C_2, ..., C_7, are the regression coefficients, FE is the fibre breaking elongation in percentage, and SFC is the short fibre content as measured by AFIS. The regression coefficients are dependent on the scales of measurement, and therefore cannot be used as a measure of their importance. To overcome this problem, 'β' coefficients of the variables are determined using the standardised variables in the regression equation. The relative contribution of the ith fibre property can be determined by the following equation:

$$C_i\% = 100 \left(\frac{B_i}{\sum\limits_{i=1}^{N} B_i} \right) R^2$$

where, B_i is the 'β' coefficient of the ith variable, N is the number of HVI fibre properties and R^2 is the coefficient of determination.

The reference set is expressed in terms of the average and standard deviation of a fibre property. In the next step, the relative difference factor for each cotton fibre is determined by the following equation:

$$D_i = \frac{(x_i - \mu_i)}{\sigma_i}$$

where, x_i is the ith fibre property of a cotton, μ_i and σ_i are the overall average and standard deviation of all the cottons in the ith property.

Now, based on the percentage contribution of fibre property $C_j\%$ and the difference factor D_j, the premium-discount index (PDI) could be calculated using the following equation:

$$\text{PDI} = \sum_{}^{N} (C_i.D_i)$$

Here the sign of the product in the summation should follow the sign of the variable as obtained in the regression equation.

2.2   Bale management

Bale management is a process to mix fibre homogeneously to get consistent production and quality of yarn and inventory control and selection of fibres according to its properties. According to the fibre characteristics bale management refers to a choice of cotton bales in order to achieve acceptable and a constant yarn quality and economical processing conditions.

Bale management is based on the categorising of cotton bales according to their fibre quality properties. The measurement of the fibre characteristics with reference to each individual bale, the separation of bales into classes, the determination of the mixture proportions and the laying-down of balanced bale mixes based on classes are included. The bale management should be undertaken because of a considerable variation in the fibre properties from one bale to another, even within the same delivery. This variation will result in yarn quality variation if the bales are mixed uncontrolled.

Bale management should be based on four properties of cotton fibres, viz.:

1. Micronaire,
2. Maturity,
3. +b value,
4. Rd value.

100% bales checking are needed for controlling above four properties in cotton mixing. Cotton bales with Rd values greater than 70 and +b value less than 9.5 is preferred than Rd value less than 70 and +b value more than 9.5 for controlling barre.

In any cotton lot mixing, whatever be the yarn lot size, the average micronaire should be kept within a range of ±0.2 and bale difference i.e. (maximum–minimum) range should be 0.6; likewise, for +b value, average range should be ±0.05 and (maximum–minimum) range should be 1.1. For Rd value, average value range should be 1.0 and (maximum–minimum) range should be 2.5. Percentage maturity of cotton should be kept 80% and above.

To achieve the above status, mixing should be done with one variety of cotton like J-34 100%, S-6 100%, Mech-1 100%, etc. with 7–8 lots at one time mixing. In no case any two of them should be mixed.

Objectives of bale management:
1. To get uniform yarn quality.
2. To minimise shade variation of the finished fabric.
3. To reduce or control fabric barre.

Flow chart of bale management:

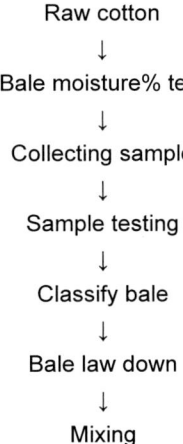

Raw cotton
↓
Bale moisture% test
↓
Collecting sample
↓
Sample testing
↓
Classify bale
↓
Bale law down
↓
Mixing

2.2.1   Observations and suggestions for improving the bale management

2.2.1.1   *Ensure cotton mixing ratio*

It is observed that if cotton mixing ratio is not maintained properly in the mixing, the controls are not so much defined resulting in inconsistency in the mixing. The following are the guidelines to ensure the cotton mixing ratio:

(i) If the cotton received is from different ginners, it is better to maintain the percentage of cotton from different ginner's throughout the lot, even though the type of cotton is same.

(ii) If different kinds of imported cotton are being used for mixing in different percentages (on weight basis), this ratio should be maintained using weighing balance.

(iii) If bales are opened manually, then, mixing ratio percentage is not ensured. Because some bales opened early while others still to be opened. It is necessary that all bales being opened should finish at same time for having mixing consistency.

(iv) To get homogeneous mixing, same percentage of each grade in each line of lay down should be maintained.

2.2.1.2   Cotton micronaire

The following are the guidelines that should be followed while preparing for daily mixing:

(i) Micronaire range of the cotton bales used should be same for all the daily mixings.

(ii) It is always advisable to use a mixing with very low micronaire range, preferably 0.6.

(iii) Consistency in micronaire must be maintained. Variation of average micronnaire of daily mixing should not be more than 0.2.

2.2.1.3   Cotton colour

(i) Range of colour of cotton bales used should be consistent for all the mixings of a lot. Range (maximum–minimum) for Rd value should be 2.5.

(ii) There should not be much variation in average cotton colour of bales used for all the mixings of a lot which should not be more than 1.0.

2.2.1.4   Pre-cleaning of bales

Sometimes ginners use metal strips to hold the cotton bales. These metal strips cause the surface of the bales to be rusty. This rust causes holes at fabric stage. Therefore, it is good to clean the bales before opening in tufts. A brush may be used to clean the surface of the bales and remove rust from the surface. Another important observation in this context is that when bales are brought to the mixing area, sometimes the wire is cut just before putting them in the lay down. Therefore, bales do not get enough time to acclimatise. It is important to cut the bale wires and clean the bales immediately when bales are brought to mixing room. This will give the bales enough time to acclimatise.

2.3   Sandwich/stack mixing

Sandwich/stack mixing is a widely used way of doing the mixing compared to automatic bale opener (Blendomat). If stack mixing is done perfectly it is a good way of mixing. Improper stack mixing will lead to barre or shade variation problem.

2.3.1   Guidelines for stack mixing

It is observed that most of the mills are using sandwich mixing method but do not control it properly which in turn results in improper mixing. The following are the general guidelines for proper stack mixing:

  (i) Mixing should be made in layers and waste should be mixed properly in the layers.

 (ii) In case of manual opening, one man–one bale at a time should be followed and worker should store all the jute, cloth, nylon or any other contamination in a bag. The common practice of throwing on the floor should be avoided.

(iii) Mixing heap should be stored at least 24 h before use; fresh mixing will cause variation and hence rejection in lap and also lap licking (in case of lap feed system).

 (iv) Bale arrangement in the lay down should be made properly. Care should be taken to ensure that two bales of same area are not placed together in the lay down to facilitate the homogeneous mixing.

  (v) In case of sandwich mixing, if cutting of mixing is not straight then it creates non homogeneity of material. It is a common mistake made in the mixing. Care should be taken for straight cutting of mixing heap in order to achieve homogeneity.

 (vi) If there is no space between wall and mixing heap, it affects the conditioning of material near the wall due to improper air circulation. To achieve homogeneous conditioning of the material, there should be some space between wall and mixing heap.

2.4   Guidelines for mixing with automatic bale opener

If automatic bale opening machine is used, the bales should be arranged as follows:

  (i) If 50 bales are used in the mixing and assuming there are five different micronaires and five different colours in the mixing, 5–10 groups should be made by grouping the bales so that each group will have average micronaire and average colour as that of the overall mixing. The position of a bale for micronaire and colour should be fixed for the group and it should repeat in the same order for all the groups.

 (ii) It is observed that safety rope around the Blendomat is not being used in some mills. It is advised that safety rope around the Blendomat should always be used and also, safety rope on both ends should be placed. It is very important from safety point of view.

(iii)   Another important point is controlling the air suction at the Blendomat. It is observed that sometimes air suction seals are not properly intact which causes weak suction.

2.5   Useable waste

During the different stages of spinning, waste is generated. The waste generated can be reused in mixing to save the cost. But there should be a balance. Using more reusable waste in mixing will result in processing difficulties which will increase the cost rather than saving. It is observed that in many mills useable waste percentage is not controlled or even not documented. This results in the inconsistency in mixing and affects the yarn quality.

It is recommended that random and uncontrolled amount of reusable waste in the mixing should be avoided at all costs since considerable count variations will result together with quality variations. It is advisable to restrict reusable waste to 5% in mixing for carded yarns and 2.5% in the mixing for combed yarns. It is also observed that in many mills, there was no department wise record of reusable waste. It is advised to maintain the record of reusable waste from each department so that it can be monitored and controlled effectively.

Identification of useable waste should be proper. If there is no identification for useable waste, there is chance of mixing the useable waste with other type of cotton. Therefore, identification is required from quality point of view.

2.6   Filling of plucker

It is observed that sometimes plucker is being filled in running condition which is not good from quality point of view. This practice causes weight variations. It is advised that plucker should be filled in idle condition and cotton inside the plucker should be pressed to avoid weight variation problems.

2.7   Different production of pluckers

It is observed that sometimes production is not same on all the pluckers that affects mixing ratio of material coming from all the pluckers. This means that material in all the pluckers is not finished at same time. In order to get homogeneous mixing; all pluckers should have same production.

2.8   Atmospheric conditions

Cotton is a hygroscopic material, hence it easily adapts to the atmospheric air conditions. Air temperature inside the mixing room and blow room area should be more than 25°C and the relative humidity (RH%) should be around 45%, because high moisture in the fibre leads to poor cleaning and dryness in the fibre leads to fibre damage which ultimately reduces the spinnability of cotton.

2.9   Cotton bale management – EFS® system software

Cotton Incorporated's Engineered Fibre Selection® (EFS® System) has earned the reputation of providing mills and merchants with the ability to rapidly process massive quantities of High Volume Instrument (HVI) data. This feature enables cotton to be selected so that all-important HVI measurements can be taken into account through the active control of averages, and statistical distributions of selected inventories of cotton bales.

Such control is economically important because cotton cost and related mill qualities, as well as processing efficiencies and associated costs, can be positively affected when cotton is acquired and used with the benefit of HVI data.

The USDA, in 1991, began HVI testing every bale of the United States cotton crop. In subsequent years, the entire domestic United States textile industry has moved to purchasing and consuming cotton using USDA HVI data. The bottom line is that of all the cotton grown worldwide only United States cotton has HVI data for every bale provided by an independent government agency, the USDA.

The availability of USDA HVI data and Cotton Incorporated's EFS® System to both domestic United States mills and selected international mills have enabled them to effectively practice what has become known as scientific fibre to yarn/fabric engineering. This engineering approach is utilised to produce yarn/fabric at optimum quality and costs consistently over time.

Scientific fibre to yarn/fabric engineering can be undertaken using Cotton Incorporated's EFS®– MILLNet in a series of steps as follows:
1. Determination of cotton specifications.
2. Opening line configuration and availability.
3. In-house inventory management.
4. Mix profile(s).
5. Bale selection.
6. Mix evaluation and performance verification.

Each of the components listed above will be briefly reviewed in order to clarify the importance of HVI data to cost-efficient raw cotton management and mill and product performance. Benefits experienced by mills using the EFS® System are listed after the conclusion of the reviews listed above.

Benefits

The following are the benefits most often reported by EFS®– MILLNet users:

1. Use of USDA HVI data eliminates cotton bale sample cutting and classing at the mill.
2. Reduction of inventory carried by the mill.
3. Just-in-time delivery of cotton improves quality.
4. Yarn quality is improved including yarn count variation, strength and Uster statistics.
5. Fewer fabric defects.
6. Elimination of cotton mix selection as a cause of barre.
7. Reduction of comber noils without loss of quality.
8. Improved efficiencies lead to lower labour costs.
9. Improved warehouse management.
10. Short fibre control improved.
11. Adjustment of mix averages and distribution %CV's based on values of incoming but not yet received cotton.
12. Better contract management, reporting and improved communications/ understanding between textile mills and their cotton suppliers.

In addition, the EFS® – MILLNet program monitors cotton contracts and manages up to 100 warehouses and up to 99 mill opening lines. The system is capable of receiving all HVI data and related cotton transaction information electronically, thus eliminating most if not all routine key punching. Warehouses are managed using portable bar code terminals. Both batch and RF units made by symbol are supported.

2.10   References

1. USTER News bulletin, Measurement of the quality characteristics of cotton fibre, **38**, 23–31, 1991.

2. Practice-of-Cotton-Fibre-Selection-for-Optimum-Mixing,   https://www.scribd.com/doc/13393448/.

3. El Mogazhy, Y.E. and Gowayed, Y., Theory and practice of cotton fibre selection, Part I: Fibre selection techniques and bale picking algorithms, Text. Res. J., **65**(1), 32–40, 1995.

4. El Mogazhy, Y. E. and Gowayed, Y., Theory and practice of cotton fibre selection, Part II:Sources of cotton mix variability and critical factors affecting it, Text. Res. J., **65**(2), 75–84, 1995.

5. Bezdudnyi, F., Planning the most economic blend in cotton spinning, Technol. Text. Ind.,**5**,3–7, 1965.

6. Ram et al., Take the guess work out of blending, Textile Ind., **128** (2), 75–77, 1864.

7. Bland, R., New finite pivoting rules for the simplex method, math, Oper. Res, **2**,103–107, 1977.

8. Murtagh, B.A, *Advanced Linear Programming*, McGraw-Hill Inc, 1981.

9. Ozan, T.M, *Applied Linear Programming for Production and Engineering Management*, Prentice-Hall, NJ, 1986.

10. SAS/OR User's Guide Version 6, SAS Institute Inc., Cary, 1989,pp. 227–316.

11. Chapter 6, *Process Control in Textile Manufacturing*, Woodhead Publishing Limited, 2013, p. 134.

12. Arbel, A. and Orgler,Y.E., An application of the AHP to bank strategic planning: The mergers and acquisitions process, Eur.J. Oper. Res., **48**, 27–37, 1990.

13. Lai, V.S, Wong, B. K. and Cheung, W., Group decision making in a multiple criteria environment: A case using the AHP in software selection, Eur. J. Oper. Res., **137**, 134–144, 2002.

14. Drake, P.R., Using the analytic hierarchy process in engineering education, Int. J. Eng. Ed., **14**(3), 191–196, 1998.

15. Korpela, J., and Tuominen, M., Benchmarking logistic performance with an application of the analytic hierarchy process,IEEE Trans. Eng.Manag.,**43**(3), 323–333, 1996.

16. Sreenivasa Murthy, H.V. and Samanta, S.K., A fresh look at fibre quality index, Indian Text.J., **111**(3), 29, 2000.

17. Hongwei, Z., Relationship of cotton fibre HVI properties with strength of rotor yarns, Int. Text. Bull., 1, 44, 2003.

18. Guha, A., Application of Artificial Neural Networks for Prediction of Yarn Properties and Process Parameters, Ph.D. Thesis, IIT Delhi, 2001.

19. Kang, B.C., Park, S.W., Koo, Y.J. and Jeong, S.H., A simplified optimization in cotton bale selection and lay down, Fibres Polymers, **1** (1), 55–58, 2000.

20. Lord, E., *Manual of Cotton Spinning: The Characteristics of Raw Cotton*, The Textile Institute, Manchester and London, 1961, pp. 310–311.

21. Norms for the Spinning Mills, The South Indian Textile Research Association, 1995, p. 1.17.

22. Application Handbook of USTER HVI SPECTRUM, Zellweger Uster, 1999, pp. 1.1–1.9.

23. El Mogazhy, Y.E., Broughton, R. and Lynch, W.K., A statistical approach for determining the technological value of cotton using HVI fibre properties, Text. Res. J., **60** (9), 495–500, 1990.

3

Checking and control

Checking and control of the key parameters at different stages of processing right from incoming raw materials to outgoing finished product is extremely important for quality assurance.

In order to have best results, it is needed to identify the key result areas (KRA) in each process and monitor them. This involves checking and controlling. We need to identify the areas that are to be checked and controlled to ensure proper quality. The control points if controlled properly should lead to the achievement of the result in the key result area, and finally the company objectives and goals. The check points are process related, whereas the control points are result related.

The normal control points in a production process are selection of raw materials, process parameters, selection and training of employees, maintenance of machines, rejection rates, delivery schedule, inventories, etc. For each control point, we can identify some check points.

The check points should be written in short and clear wordings so that even a less literate person could also understand and implement without waiting for further instructions. It is essential to impart required training to the people, assigned the job of checking, so that there shall be no bias. People on spot should get reliable inputs to take decisions relating to the corrective and preventive actions depending on the findings while conducting checks. The man who conducts the checks should be clear about the purpose of that check, so that he can be precise while making observations. If the objectives are not clear, there are dangers of just entering some data without proper verification. By this the decision or action taken shall be futile [1].

Better yarn properties do not necessarily involve in increasing costs, but in the long-term they usually provide a reduction in costs. The following possibilities are introduced by some investigators [2–17]:

- Detection of raw material variation and its reduction (a basic and fundamental requirement).
- Reduction in the coefficient of variation values of the yarn quality characteristics (primary requirement).
- Elimination of rare yarn faults (weak places), which is becoming more and more important.

- Increase of the yarn breaking force and elongation at break.
- Application of high-quality preparation machines and spinning units (or improvement of those available).
- Supervision and control at all the spinning processes (also a basic requirement).
- Obtaining optimum conditions with all machines.
- Use of automation (according to existing mill conditions).
- Elimination of any negative ambient influences.

Many research works have been focused on yarn engineering [8, 9, 14, 15, 18–24]. According to the literature, yarn engineering refers to the following:

- Obtaining optimum conditions in terms of product quality, with respect to the yarn and the end product.
- Optimum selection of the raw material for the required quality.
- Increase of the added value by means of a better use of the raw material.
- Pre-determination of the yarn properties based on raw material and process data, i.e., the quality data of the yarn and the end product are no longer to be considered as chance conditions.
- Ensuring the quality level throughout the complete process.
- Keeping constant quality in order to ensure long-term marketing conditions.
- Reduction of manufacturing costs by increasing efficiency.

With process management, not only is each individual machine in the spinning mill to be run under optimum conditions, but also each separate processing stage is to be exactly tuned to the other processing stages in order that a reasonable and process-oriented compromise can be arranged with respect to quality and costs [8,18–24]. For this purpose, the following is necessary:

- Testing of the fibre properties before and after each important processing stage.
- Correct settings, in order to achieve optimum conditions at all machines, taking into consideration the yarn as the end product.
- Determination of the most suitable machine equipment.
- Arranging optimum conditions for machine maintenance, so that there is no reduction in quality as a result of long-term running of the machine.
- Introduction of early warning systems.

3.1   Checking at different stages of a cotton spinning mill having 25,000 spindles is given below

Stages	Types of studies	Frequency of checking	Remarks
In-coming (cotton bales)	Cotton fibres properties checking –100% bales are to be covered	Daily	Four parameters are to be considered for bale management, viz. (i) +b value, (ii) Rd, (iii) Micronaire and (iv) Maturity. This is very much needed for controlling barriness in the knitted fabric.
	Sugar content in raw cotton	As and when required	Excess sugar content in raw cotton gives stickiness during processing (to be discussed later on)
In process mixing	Manually picked up cotton are checked to see any presence of foreign matters (For export purpose)	Daily: Picker wise by name	This is needed to keep a check on manual hand pickers engaged to pick up foreign matters i.e. contaminants from cotton bales
Blow-room	Blow-room waste % beater wise and overall	Fortnightly	Overall waste% should be approximately 1.0–1.5% more than trash% in raw cotton. Higher the trash% in cotton, more will be waste%
	Blow-room-cleaning efficiency%	Fortnightly	It should not go up more than 65%. More the trash% in cotton, more will be the CE%
	Lint% in blow-room droppings	Fortnightly	Lint% should not be more than 35% (tested by trash analyser)
	All beaters speed	After mixing or speed change	This is particularly required when trashy cotton or less matured cottons are processed
	Beater wise testing of cotton fibre's properties either through AFIS or SPIN LAB. AFIS is preferred (testing of short fibre percentage, neps, mean length, etc.)	As and when required	This testing helps to see any adverse effect of beaters on cotton
Carding	Neps/g of carded slivers from each card	Two times in a week particularly after mixing change	Maximum value of neps/g is 50
	Cleaning efficiency of each card	Fortnightly or after mixing change	CE% should not be less than 95%
	Trash% in card slivers from each card	Fortnightly or after mixing change	Maximum value of trash% in card slivers should be 0.18
	Total waste% from each card	Daily 3–4 cards	Total card waste/should be kept in between 5% and 6%
	Testing of card sliver and corresponding feed materials through AFIS. This is to know whether best card performance is achieved or not	Once in a week or after mixing change to cover all cards	Individual cards performance is known and achieved
	Speed of each component of card like – taker-in, cylinder, doffer, flat	As and when required	This is needed for standardisation in processing
	U% from each card	Two times in a week covering all cards	This is needed to see whether any card is giving very high U% or not

Contd...

Contd...

Stages	Types of studies	Frequency of checking	Remarks
	Card sliver 100 yd wrapping from each card	Alternate day or two times in a week	CV% should be maximum – 1.0% (This standard is applied for chute feed card attached with long term auto-leveller)
Drawing	U% of breaker and finisher	Daily	Standard U% should be fixed for breaker and finisher
	5 Yd wrapping of each drawing is to be taken –2 readings/delivery	Daily	A standard sliver hank is to be fixed up for both breaker and finisher drawing
	Spectrogram of each drawing is to be taken on rotation basis	All drawing should be covered in a week and faster rotation is to be done	This will help to control periodic variation and drafting wave. Both of these faults have an adverse impact on fabric if present

Following checking is needed for RSB – 851 drawing with auto-leveller

Stages	Types of studies	Frequency of checking	Remarks
	Optimisation of levelling time checking (S-93 setting)	Fortnightly or as and when required	Keeping running setting at 5.0, spectrogram to be taken at settings 3, 4, 5, 6 & 7 i.e. two digits ahead and two digits behind of running setting. If wave-length comes in between 30 and 70 cm, then setting is wrong, otherwise ok
	To check efficiency of levelling	All drawings are to be completed within a week and then further rotation is done	From a single full drawing can, 50 readings are to be taken each from three different position of the can i.e. top, middle and bottom position for cut lengths of 2, 3 and 5 m. If following CV% of wrapping is found, then it is ok. 5 m cut length CV%: 0.30 3 m cut length CV%: 0.50 2 m cut length CV%: 0.60
	Adaptation knob setting (B-90)	As and when required, particularly when wrapping variation comes	First the mechanical draft is to be adjusted and auto-leveller to be put off. Then wrapping is checked. For finer setting, auto-leveller is to be switched on and the adaptation key to be adjusted which lies between 300 and 700
	Trumpet and scanning roller adjustment for RSB-851 For RSB-951, trumpet change is not needed	When sliver hank is changed	To be checked whether the following adjustments are maintained or not Standard hank / Trumpet (mm) / Scanning roller (mm) Up to 0.100 / 4.2 / 6.4 0.101–0.150 / 3.8 / 6.4 Above 0.150 / 0.8 / 4.9
	When the wrapping fluctuation occurs very frequently then 8-setting adjustments by calibration meter is needed	Once in a month through electronic department	Calibration meter is supplied by the drawing manufacturer. The process of checking is discussed on later stage
Simplex	Full doff simplex breakages from each simplex in the following fashion: a. Start to 1/4th doff position b. ¼th position to ½ position c. ½ position to ¾th position d. ¾th position to full position Based on above breakages, full doff breakages to be calculated	Daily two simplex machines	Full doff breakages should not be more than 1.5% (breakages/100 spindle hour) If first layer breakage is high, then winding wheel arrangement is wrong. If breakage steadily go on increasing then tension is high i.e. rachet wheel adjustment is wrong. If there is any bursting of roving during spindle rotation, then tpi is less

Contd...

Contd...

Stages	Types of studies	Frequency of checking	Remarks
	U% of simplex bobbins from each simplex. Two bobbins – Front row. Two bobbins – Back row	Daily all simplex m/c's	Standard norm of *U*% is to be fixed by the mills
	5 yd wrapping of each simplex. Two bobbins/ simplex Total readings – 8.0	Daily all simplex machines	It is better to fix up roving length as per draft of ring frame. Suppose ring frame draft is 30, then 120 yd (one lea) ÷ 30 = 4 yd of roving length is to be taken for wrapping

	Stretch % of roving 6 simplex bobbins are to be taken
	          Front Back
	          row row
	 Gear end - 1 1
	 Middle - 1 1
	 Off end - 1 1
		Minimum one simplex machine per day	Stretch% is given by:
			 $$\frac{\text{Wrapping at full doff} - \text{Wrapping at start}}{\text{Wrapping at start}} \times 100$$
			 Maximum stretch% = 1.0

	Spectogram analysis of each simplex	To be taken while taking *U*% of the bobbin	This checking will help to curb the periodic faults and drafting wave in the roving
	60 yd lea strength of roving taking 10 bobbins per machine for checking the tpi adjustment in simplex	While changing roving hank and lot change	60 yd lea strength should be 30–35 lbs. If this level is achieved, then tpi in roving is OK. Higher value mean that tpi is to be reduced and vice-versa
	Take 30 simplex bobbins and calculate wrapping CV. Values of 60 reading of 5 yd length of roving is to be taken	One simplex per day daily	Maximum wrapping CV% should be 1.5%
Ring frame	Ring frame wrapping is to be taken daily and also lea strength for CSP value. Eight bobbins each from RHS and LHS (1008 spindle) – total 16 bobbins	Daily covering all ring frames	Count wise CV%, CSP and lea strength CV% are to be calculated. This will reflect over all CV%. For count CV% standard is 1.80–2.0. For lea strength CV% it is 5.0–6.0
	Spindle speed is to be checked by stroboscope to find out low spindle/ bobbin speed and hence low tpi	Six machines per day on both sides of ring frame	This checking is needed to control low tpi in yarn
	U% checking – Four bobbin/ring frame to be taken and all the ring frames to be covered daily. Care should be taken not to repeat same spindle. It is better to keep a register for not repeating spindle number	Daily to cover all ring frames. If not possible then at least 10 m/c's per day on count basis	Rogue spindles are to be identified for action
	tpi checking just after count change and to find out tpi CV%. 20 readings are to be taken from 10 ring frame bobbins	After count change or at least two ring frames per day	tpi CV% should be 6% maximum

Contd...

Contd...

Stages	Types of studies	Frequency of checking	Remarks
	Ring frame breakages in the following manner: (i) Bottom at slow speed (ii) During speed change from slow to high (iii) Middle breakages – 1 h (iv) Top breakages – 20 min	Two ring frames/day	Average full doff breakages should be maximum 2.0–2.5%
	Start-up breakages in ring frame	On complaint from production department	It should be maximum 7–8%. Higher start-up breaks contributes for more gaiting up time and more pneumafil waste
	Checking pneumafil suction pressure by ATIRA Suction Pressure Instrument(ASPI)	To cover all ring frames in a month	Suction pressure for 1008 spindles ring frame should be: Pneumafil end – 15–16 middle of m/c – 13–14 off end – 9–10
	Variance length curve of 10 ring frame bobbins from individual ring frame	No definite schedule for checking	If 4 bobbins out of 10 bobbins i.e. 40% readings show the same trend of deviation, then only proper investigation is needed. Repeated testing is needed for the verification of the same trend. Frequent checking help to arrive at the definite idea for action area
Ring doubling	Ply yarn tpi checking – 10 bobbins × 2 readings per bobbin i.e. total 20 reading	On creeling a fresh and new lot on ring doubling machine	To follow the tpi standard i.e. Actual tpi should be 2–3% less than mechanical tpi
	From 10 ring doubling bobbin, 20 readings are to be taken for checking count, CSP and corresponding CV%	From each lot – one time	Count CV% should be 0.8–1.0% and lea strength CV% should be 3–4%
	U% and imperfections checking from 10 ring doubling bobbin	From each lot-one time	Standard norms are to be followed which are fixed for the lot
Cone winding	To check the winding breakages of 50 ring frame bobbins and to collect the faulty portions for fault analysis	Each lot is to be covered. Frequency of this study depends on lot size	Analysis of yarn faults gives the opportunity to improve the yarn quality in the back process
	Bobbins to cones – U% and imperfections are to be checked from auto coner to see increase in hairiness and imperfection. Six cones and corresponding six ring frame bobbins are to be taken	New lot coming to savio plus running lots on savio. Frequency of checking depends on lot size	Lot size No. of checking 1 Ton - 1 2 Tons - 2 3 Tons - 2 4 Tons and above - 3 Maximum increase in imperfection should be 30–35%. Mainly neps are increased. Increase in hairiness usually go up by 1 unit from ring frame value
Outgoing	From each lot, 30 cones are to be taken for checking count and CSP and corresponding CV%(For single)	Frequency of checking depends on lot size like above	Count CV% and lea strength CV% depends on the standard fixed up by the Mills. For universal acceptance, count CV% should be 1.2–1.3% and lea strength CV% should be 5–6%

Contd...

Contd...

Stages	Types of studies	Frequency of checking	Remarks
	From each lot, 10 full cones are to be taken randomly for checking single yarn RKM	20 reading per cone. Total 200 reading for each lot	**RKM(g/tex)** Carded single hosiery yarn –13.0–13.5 Carded single warp for weaving –15.0–16.0 Combed single yarn –Carded single hosiery yarn + 1.5 **2 ply:** Carded hosiery (tpi – 60–65% of single yarn) Carded hosiery single yarn + 2–3 g/tex Warp yarn(TPI –80–90% of single yarn) Carded single warp yarn + 4–5 g/tex
	30 Cones are to be taken for rewinding and cutting yarn faults in a manual winder. Manual winder's eye setting will be closer than Auto-Winder's Eye setting	Frequency of checking depends on lot size Lot size (ton) No. of studies 1.0 1 2.0–3.0 2 4.0–6.0 3	Rewinding faults per 100 km should be 1.0 or less than 1.0 for export market and 2.0–2.5 for domestic market
	If classimat is available, then 5–6 cones are to be taken whose total meterage should be more than 100 km	One time for each lot	Faults allowed to remain in final cones are usually chosen by the buyer. If not given, following number of faults are to be taken as standard for 100 km of yarn a. Objectionable faults – 0.5–1.0 (A4 + B4 + C4 + D4 + C3 + D3) b. Serious faults – 3.0 (A3 + B3 + C2 + D2) c. Long thick (E+G) – Less than 2.0 d. Long thin (H+I) – less than 6.0 (Above parameter are for export quality) Domestic quality seldom requires classimat faults
	Moisture regain% from cones by moisture meter where xorella type conditioning machine is not available. All cones are to be kept in conditioning room for 24 h far attaining a MR% value more than 7.5% before final packing	Daily from each lot before packing from 10 cones	This is very important otherwise there will be an invisible loss and yarn realisation will be less due to less moisture
	Packed cartons weight checking	Each day from each lot randomly 10 cartons are to be taken per lot	This checking is necessary to monitor packer's alertness while packing cones
Comber checking			
Comber	Ribbon lap meter to meter CV% checking and g/m checking	Twice in a week	Full length lap CV% should be 1.0–1.2% standard g/m is to be maintained
	Noil% checking delivery wise	Twice in a week	Head to head difference should not exceed 1.5% (max–min)
	Combed sliver *U*%	Alternate day	*U*% should be within standard norm
	Neps in combed sliver	Twice in a week	Neps/g should not be more than 20
	Trash% in combed sliver	Once in a week	It should be 0.05–0.06%

3.2　　How to check sugar content in cotton

Honey dew is the main cause of sugar content in the raw cotton which in turn manifest in stickiness. Stickiness results from physiology of the plant and from insect secretion (aphides and honey dew).

During ginning, honey dew droplets spread over the entire fibres in the small droplets form, which is difficult to detect by naked eyes.

Stickiness is generally measured by a number of techniques as:

1. Indirect measurement of sugar content in the cotton.
2. Minicard method.
3. Thermodetector method.

3.2.1　　Sugar content checking by indirect method (by Benedict's reagent) [25]

Procedure: About 1 g of cotton selected randomly (free from foreign matters and impurities) is boiled for about 5 min in 30 ml of Benedict's reagent solution, which is prepared as follows:

10 ml of Benedict's reagent + 20 ml of distilled water i.e. 30 ml of diluted Benedict's reagent is taken for 1 g of cotton.

After cooling the colour of the boiled solution is compared with standard colour displayed on Benedict's bottle. Sugar content is observed as follows:

Colour of the boiled solution	Sugar content in raw cotton (ppm)
Blue	0.25%
Green	0.50%
Yellow	1.0%
Red	2.0%

Normal sugar content in cotton is 0.3%.

3.2.2　　Minicard method [26]

The physical interaction of all sugars on lint with equipment can be measured by several types of machines. The primary difficulty with these physical tests is in standardising the stickiness measurement. As with chemical testing, these tests must be correlated with measures of fibre processing efficiency in order to interpret the results. One of these tests, the minicard, is a physical test that measures actual cotton stickiness of the card web passing between stainless steel delivery rollers of a miniature carding machine. Modelled after a production carding machine, the minicard must be run under strict tolerances. A '0' minicard rating indicates that no sticking was observed, while progressively higher numbers (on a 0–3 scale) indicate progressively greater amounts of sticking during the process. Cottons with high plant sugar contents evenly distributed along the fibres may fail to be measured as sticky in this test. The minicard test is slow and has been replaced as the international standard by the manual thermodetector.

3.2.3   Thermodetector method [27]

The thermodetector is the best instrument to check each bale effectively. In the thermodetector, the test sample (2.5 g) prepared in the form of a web (54 cm × 16 cm) is sandwiched between two sheets of aluminium and the preparation is subjected to pressure at 80° C for 12 s (the heating element being in contact with the upper aluminium sheet). The prepared sample is then removed and placed on a table for at least 30 min before counting the sticking points. After removal of the web, the sticking potential of cotton is determined by visually counting the number of fibre points stuck to the upper and lower aluminium sheet.

In the latest generation thermodetector, counting of sticky points is done with a video camera by scanning technique, and the image is analysed by the computer.

3.3   Checking of RSB – 851 calibration by calibration meter

There are eight checking points enlisted in the calibration meter which is supplied by RSB drawing manufacturers.

1st Checking: Scan rollers voltage checking is made by three numbers of feelers gauge as 3, 4 and 5 mm. Connect the calibration meter by pressing the top socket of the meter on the designated point of the levelling unit. Make B-90 adjustment to 500 before checking and keep this till eight checking points are checked. Now insert the above feeler gauge one by one and press the pressure handle of the scan rollers on the press position and check voltage on the calibration meter.

At 3 mm feeler gauge, voltage reading   –   3.3 V ± .01

At 4 mm feeler gauge, voltage reading   –   5.0 V ± .02

At 5 mm feeler gauge, voltage reading   –   6.7 V ± .01

If the above voltage is not found, then either scan roller setting is disturbed or scan roller is worn out.

2nd Checking: Motor brake – For this checking, move the motor belt by hand and observe that motor runs for few seconds with a click sound and then stop with a click sound. As we drive the motor by hand, this gets repeated. Motor cover is to be kept opened to observe that the motor is running. If the motor does not run, then the motor is to be taken out for checking or problem may be with differential gears.

3rd Checking: Zero drifting – At the zero position of DC motor, there will be no movement in motor. For this checking, connect the ground connection of the meter to connector strip at no.38 position (Last position). Keep Q-90 switch on. This direction has been given in the calibration meter. Put on the switch at 3rd position on the meter. Motor will not run. If motor runs, then the setting is disturbed. For this checking while the drawing machine is decreeled, take out earthing connection from E-3 of N-III panel.

4th Checking: Tacho generation – With the machine in running condition and 4th position switch 'ON' in calibration meter, the voltage in the calibration meter should come in between 19.5 and 20.0 V. If it is not shown then there is problem with the carbon bush of the motor.

5th Checking: As per direction shown in the calibration meter whether to run the machine or stop the machine, switch-on the 5th position and meter reading should come as $0.7 \times$ main draft. If it is not shown, then main draft calculation is wrong.

6th Checking: For this checking, reading of S-93 switch is to be noted down. Keeping it at 5.0 and with the 6th position "Switch-on", the meter reading should be 192-5–187. During this checking, whether the machine will run or stop, this guidance is given on the meter. 'O' indicates stop and 'I' indicates run as shown by the meter.

7th Checking: There are five numbers of red lights on both sides of green lights on material feed position. Each light corresponds to 5% load. When five lights are 'ON' position it means 25% load. When tester shows +27.5% load, the machine stops. Again with +22.5% and−22.5% load, the machine does not stop. If machine stops at a load of +22.5% and−22.5%, then there is either electrical problems or machine is jammed.

8th Checking: Before this checking, keep R-38 position on 5. (If it is not at 5, then make it at 5) Switch on the 8th position switch on the meter and follow the instruction given in the meter to see whether the machine is to be run or stopped. The meter reading will come 1st as 1.35 and the 0.79.

3.4   To minimise the start-up (breaks) in ring frame

Following action are found effective to control start-up breaks in a ring frame:

1. To see that under winding coil should get properly trapped by new bobbin.

2. Start-up breaks may be reduced if lappet height at bottom position is increased by 5–6 mm.
3. By increasing the under winding coils laying time by 2–3 s.

3.5   Contribution of post-spinning departments on quality

Post-spinning department now-a-days is considered very important where technically qualified persons are needed for technical supervision. Previously non-qualified persons were used to be employed (till now many mills do so) with the idea that there was no need of any technical skill in post-spinning department. But this is a very wrong concept. There is a big technological upliftment in the manufacturing of post-spinning machines, particularly Autowinder. Autoconer machine with inadequate technical supervision and attended by unskilled winder can give rise to various types of market complaints. Here, certain important areas of Autowinder (Savio-Esperol) which are worth to be noted to achieve maximum benefit of Autowinder are described. These areas are:

1. Setting of proper cone formation.
2. Implement of VSS.
3. Importance for checking (cycles/splice) ratio.
4. To check Winder's efficiency and savio efficiency.
5. Count adjustment group wise or lot wise.
6. Impact of unskilled worker on winding.
7. How to avoid extra yarn with splice.

3.5.1   Setting of proper cone formation

There are three factors which control the proper formation of cone. These are:

(a) Yarn tension
(b) Counter weight
(c) Friction clutch

Yarn tension is to be kept in between 90 and 100 mbar, counter weight in between 3.0 and3.2 bar and friction clutch to 1.1–1.2 bar (Autoconer as Savio Espero L).

For soft and low twist yarn, the cut cone problem often occurs, for which the following steps to be taken:

(a) Yarn's speed – Low.
(b) Yarn tension – Low.

(c) Counter weight – More.

(d) Friction clutch – Low.

3.5.2   Implement of VSS

VSS which means variable spindle speed should be made effective particularly for hosiery yarn. A typical setting is suggested below:

Top – '0'.

Change – 75%.

Bottom – 40%.

Yarn speed – 1200 mpm (more or less).

Meaning of the above setting adjustments are: change 75% i.e. the slow speed is to be effective when the feed cop's position comes to 75% position from the top or ¼th position. Bottom 40% means that the slow speed at the bottom i.e. from ¼th position to exhaust position should be 40% less than the yarn speed at ¼th position which is 1200 mpm as shown above.

Calculation of yarn speed:

From top to 75% position of the cop $= 1.031 \times 1200 \times \dfrac{2}{2 - 0.1}$

$= 1301$ mpm

Bottom slow speed $= 1301 \left(1 - \dfrac{40}{100}\right) = 781$ mpm

3.5.3   Importance of checking cycles to splice ratio (attempts for one splice)

This checking reveals whether any drum is giving excess cuts or false cuts or not. Cycle to splice ratio of a drum should come 15–20% higher than total splices (Total splices = Number of bobbin changes + faults cutting). If it is more than above, then there is some problem either in electronic or mechanical or it may be due to many weak places in the yarn which are causing more number of splicer's operation. This is the duty of the departmental supervisor to check each drum through computer and to identify the rogue drum for corrective action. Following example will make the picture clearer:

Suppose, two drums of any Autoconer (Savio Espero L) are showing the following reading:

Drum no.	Cycles	Bobbins	Cuts	Splices	Red light
1	39530	11389	10734	27894	1598
2	43680	10535	10819	31077	1623

Drum 1: Cycles to splice ratio = 39530/27894 = 1.41.

Drum 2: 2 Cycles to splice ratio = 43680/31077 = 1.41.

Above drums are showing an average attempts of 1.4 for one splice. It should be maximum 1.2. So, one should take these two drums for analysing the reasons for higher attempts for a splice.

3.6   How to check machine efficiency, splicer efficiency and operator efficiency

1. **Machine efficiency**: Whole machine efficiency depends on individual spindle efficiency. Those spindles having lower efficiency than average efficiency should be checked and analysed in each shift. This is the prime duty of technical supervisor.

2. **Splicer efficiency**: The cycle to splice ratio at every spindle position to be maintained less than or equal to 1.2 for better performance. Spindles should be checked for finding out reasons of higher values if any. Shift wise checking is needed.

3. **Operator efficiency**: Find out the red light time, bobbin wait time and doffing wait time from the machine's computer. Bobbin wait time should be zero. Both the Red light time and Doffing wait time should be kept as low as possible preferably below 20 s to improve the machine's efficiency.

3.7   How to adjust count lot wise in Savio

In savio autoconer, when we feed any count then electronic channel in savio takes it as ADMV. ADMV stands for "Analog Digital Mean Value". ADMV depends on count, colour of the yarn, blend, types of fibres, denier, etc.

Suppose three types of dyed fibres of same count are running in one group of savio –ADMV will be calculated based on 3 types of different coloured fibres and others parameters as stated above though count is same. Suppose, 1st lot's ADMV is 700, 2nd lot's ADMV is 750 and 3rd lot's ADMV is 731. The average of three lots is 727. This value differs from 1st and 2nd lot and as such an erratic signal value from EYC will go to the 1st and 2nd lot drums resulting in difference in cleaning efficiency which will result either false cutting or less fault in cutting of the yarn.

So, it is advisable to use different groups for different coloured lots though having same count (Autoconer taken is Savio Espero L fitted with UPM-200 EYC).

3.8   Skilled operators in Autowinder are essential

Autowinders remove the yarn's objectionable faults and serious faults making it suitable for forward process but, apart from good autowinder, manual skill to operate the machine is most urgent otherwise problems in knitting or weaving preparatory processes come. Following problems appear in the cone due to winder's negligence.

1. Bunches in cone.
2. Parallel yarn in cone.
3. Extra yarn with splice in cone.
4. Loose fluff/flies in cone.
5. Layer's slip in cone.
6. Excess hairiness in cone.

• **How bunches in cones come**

Reasons:

1. When yarns go out of the tension discs due to accumulation of fluff and flies or hard wastes in between the two discs resulting in the formation of soft layer in cones. These soft layers get sloughed off from cone during unwinding process.
2. If suction nozzle touches the cone, the layers in the cone get disturbed causing sloughing off during unwinding process.
3. If there is repeated splicer operation due to yarn's cut, then during each piecing there is a reverse movement of the cone causing top layers, in contact with the drum, disturbed and become loose. These layers during unwinding of cone get sloughed off.
4. After each break of the yarn, the cone gets lifted from the contact of the rotating drum. If the lifting process fails, then the surface of the cone gets abraded against the running drum making layers to slip during unwinding process.
5. If there is remarkable ribbon formation in cone, then layers will get slipped.

• **How parallel yarn comes in the yarn**

Reasons:

During ejection of the running bobbin from the bobbin holder, the yarn from the ejected bobbin gets cut. If this does not happen, then one end from the running bobbin which runs through the Eye-slot will run in parallel with the ejected bobbin end running out of Eye slot and thus will go on to the cone forming parallel yarn. This problem gets increased when bottom conveyor belt stop running.

• **How extra yarn with splice comes in the yarn**

Reasons:

1. Failure of scissors.
2. Wrong splice channel or J-Channel setting.

Extra yarn with splice is also called 'Y' yarn. Suppose J-Channel setting is 86% and 3 cm. 86% means that after splicing, yarn's cross section which is increased by more than 86% will be cut. J-channel closer setting is required if extra yarn with splice appears. While checking rewinding of savio cones, it should be taken with the drum number noted inside cones so that the drum number is highlighted through rewinding faults. If scissors are ineffective, the operator can detect the fault. Operator's knowledge and cooperation is very much essential in this case.

• **How loose fluffs and flies come with the yarns in cone**

While running carded cotton particularly with hosiery twist, loose fluffs and flies are generated in the department in sufficient amount. These fluffs and flies get settled on machine parts and on floor too. There should be systematic cleaning of these fluffs and flies from the machine and also from the floor by the operator. These fluffs and flies are usually cleaned by the compressed air through a long pipe. During cleaning, the Auto winder should be stopped; otherwise loose fluffs and flies get attached with the running yarn. These loose fluffs and flies generate while running cotton yarns, more particularly for carded cotton yarns with low tpi.

• **How hairiness increases in the yarn more than expected**

Sometimes, increase in hairiness from bobbins to cones is more than usual expectation. This is due to the following reasons:

1. Yarn tension adjustment is on higher side.
2. Tension discs developed scratches.
3. Accumulation of fluffs and flies in between wax roll and tension disc.
4. There are scratches on tension disc spindle.
5. Cones are getting heavily rubbed against the drums due to wrong adjustment of cone setting.
6. Yarn is touching metallic parts of the machines and thus getting abraded.
7. Timely change of wax roll is not done (for hosiery yarn only).

3.9   How to achieve good splice strength in Air Splices model-490 in Autowinder

Splice retention strength should be above 80% of the normal strength.

Following points are to be taken care of:

1. Splice introducer movement should be free and not to stop half way.
2. Chamber cover should be free from any groove.
3. Chamber cover spring should work properly.
4. Groove formation or cut in the preparator is not conducive for getting good splice strength.
5. Yarn trapper should swing properly to its optimum distance during splicing. If it swings half way, then splice appearance will not be good, though splice strength will not be affected.
6. Incoming air pressure should be sufficient as follows:

    Incoming air pressure – 6 bar,

    Operating air pressure – 5.5 bar,

    Machine should stop at 5 bar.

    For old m/c:

    Incoming air pressure – 7.0 bar,

    Operating air pressure – 6.0 bar.
7. Lever operation should be free, particularly the lever that operates the chamber cover.
8. Splice area should be cleaned once in 8 h by compressed air to clean fluff/flies.

3.10   References

1. Purushothama, B., A Practical Guide on Quality Management in Spinning, Woodhead Publishing India Pvt. Ltd., 2011, Chapter 8, p. 191.
2. Zellweger Uster®, Uster® Evenness Testing – Application Handbook, Zellweger Uster Ltd. Publication, 1986.
3. Zellweger Uster®, Uster® Statistics 1997, Uster® News Bulletin, No. 40, May 1997.
4. Zellweger Uster®, Uster® Statistics 1989, Uster® News Bulletin, No. 36, October 1989.
5. Zellweger Uster®, In Mill Training, Uster Training Center.
6. Zellweger Uster®, Uster® Tester 3, Operating Instruction Manual, August 1990.
7. Zellweger Uster®, Quality Management in the Spinning Mill, Uster® News Bulletin, No. 39, August 1993.
8. Zhu, R., et al., The prediction of cotton yarn irregularity based on the "AFIS" measurement, J. Text. Inst., 1996, **87**, Part 1, No. 3, pp. 509–512.
9. Nakamura, M., et al., The effect of bale opening on the quality of intermediates in the spinning process, J. Text. Inst., 1997, **91**, 151–164.
10. Mogahzy, Y.E., Developing Fundamental Measures of Cotton Multi-Component Blending Performance, American National Textile Center Annual Report, F99-A13, November 2000.

11. Mogahzy, Y.E., Utilization of Continuously Monitored Data in Product Improvement, American National Textile Center Annual Report, A92-C2, September 1993.

12. Cherkassky, A.E., A computer simulation of yarn breakage in the ring spinning process. Part I: model structure, investigation strategy, and experimental design, J. Text. Inst., 1997,**88**, Part 2.

13. Cherkassky, A.E., Computer simulation of yarn breakage in the ring spinning process. Part II: description and analysis of the simulation results, J. Text. Inst., 1997,**88**, Part 2.

14. Fraser, W.B., Dynamics of High Speed Yarn Transport in Textile Processes, American National Textile Center Annual Report, S92-C5, September 1993.

15. Nikos, I., et al., Production Planning and Control in Textile Industry: A Case Study, German National Research Center for Information Technology – Annual Report, 1996.

16. Koo, H.J., et al., Variance tolerancing and decomposition in short staple spinning process. Part I: modelling spun yarn strength through intrinsic components, Text. Res. J., 2001,**71** (1), pp. 1–7.

17. Koo, H.J., et al., Variance tolerancing and decomposition in short staple spinning process. Part II: Simulation and application to ring and OE spun yarns, Text. Res. J., 2001,**71** (2), pp. 105–111.

18. TCC-Netherlands, Third Technical Assistance Report for El-Minia Spinning Mill, Textile and Clothing Consultants Bv, June 1999.

19. TCC-Netherlands, Fourth Technical Assistance Report for El-Minia Spinning Mill, Textile and Clothing Consultants Bv, October 1999.

20. TCC-Netherlands, Fifth Technical Assistance Report for El-Minia Spinning Mill, Textile and Clothing Consultants Bv, January 2000.

21. TCC-Netherlands, Sixth Technical Assistance Report for El-Minia Spinning Mill, Textile and Clothing Consultants Bv, June 2000.

22. TCC-Netherlands, Seventh Technical Assistance Report for El-Minia Spinning Mill, Textile and Clothing Consultants Bv, June 2000.

23. TCC-Netherlands, Technical Assistance Report for El-Minia Spinning Mill, Textile and Clothing Consultants Bv, October 2001.

24. James de Wet, Report on Product Quality, Quality Management and Cost of Quality, CSIR Division of Textile Technology, May 13, 1993.

25. Benedict,S.R., A reagent for the detection of reducing sugars, J.Biol. Chem., 1909, **5**(6), pp. 485–487.

26. Ali, N.A. and Mohamed, B.A., A modified chemical method for cotton stickiness grading suitable for commercial application. World Cotton Research conference. 2. Athens, Greece 6–12, Sep. 1998.

27. Frydrych, R., Détermination du potentiel de collage des cotons par thermodétection, Coton Et Fibre Tropicales, 1986, **XLI**, fasc. 3, 211–214.

Spinning yarn is a process of converting fibre materials into yarn. Fibre materials are processed through a number of machines and the fibre materials are converted in steps into fibre assemblies in different forms and ultimately into yarn. For establishing an effective system to implement quality assurance in a spinning mill testing different key characteristics of fibre materials, fibre assemblies in different forms and yarns are essential.

At first when the raw cotton arrives, sample is collected from each bale and respective identification number is given to each bale. Samples are tested in HVI machine and from the obtained results; the bales are classified in suitable categories for bale management and stored in the warehouse. Raw cotton is fed to the blow room through the bale management system. In blow-room, the fibres are opened and cleaned by beaters. If over beating takes place then there is a chance of fibre damage. Tests are undertaken of cotton samples collected before and after every beating position to check the cleaning efficiency and fibre damage. Different wastes that are collected at the fibre deposition points of blow room are also studied to assure proper removal of trash particles as well as least possible wastage of cotton fibre.

In the carding machine the ktex of the produced sliver is checked once per shift. Carding performance is evaluated by testing card sliver in USTER evenness tester and AFIS. Nep removal efficiency, SFCw, SFCn, UQL, etc. are seen and necessity of gauge change, grinding and wire change of cylinder, licker-in and flat is determined. The card sliver is tested in USTER once per day to test its unevenness ($U\%$) and coefficient of variation of mass ($CV\%$).

In breaker draw frame, ktex of sliver is tested once per shift. Sliver is also tested in AFIS to determine the length of the fibres in strand to set the proper roller gauge in the roller drafting zone. The breaker sliver is tested once per shift in USTER to determine its unevenness ($U\%$) and coefficient of variation of mass ($CV\%$).

In lap former, mainly the lap thickness is measured. Performance of comber plays a vital role on the quality of the combed yarn. Nep removal efficiency of the machine is determined by testing samples collected after combing. Also the noil is tested to check wastage of long fibres during the processing. The ktex of combed sliver is checked once per shift. The combed sliver is also tested once per shift in USTER for U%, CV%, etc. and estimating periodic faults from the spectrogram.

The finisher draw frame is the last machine fitted with an auto leveller where the mass variation of sliver can be maintained. If the variation of the finisher sliver can be controlled, then there will be less variation and imperfection in the yarn. Also the calibration of the auto-leveller is done when necessary by the Electronic department. The finisher sliver is tested once per shift in USTER to determine its U%, CVm%, CV3m% and to see whether there is any periodic fault from the spectrogram.

In simplex, the hank of roving is mainly tested once per shift. Also the roving is tested in the USTER for U% and CV%. A schedule is maintained so that all the spindles of simplex machines are tested over a period of time.

The yarn of ring frame, rotor machine and autoconer are tested for yarn count. This test is carried out at each shift for each count and lot. Testing is carried out in USTER to determine U%, CV%, relative count, +50% thick places, −50% thin places, +200% Neps (ring yarn), imperfection index, hairiness, standard deviation of hairiness, periodic faults, etc. From there necessary actions are taken to eliminate or reduce the faults. The CSP (Count Strength Product) of the yarn are also tested to determine the bundle yarn strength. The TPI (twist per inch) of yarn is tested at the beginning of each lot.

The moisture content in the yarn packages, after autoclaved, are also tested by the QA department so that proper time of streaming can be maintained to deliver yarn with required amount of moisture content. Where heat setting facility is not available, cones are to be conditioned under 80–85% moisture in the conditioning room for 24 h and moisture regain% to be checked by moisture meter before packing, which should be above 7.5%.

4.1   Testing instruments

Today's spinning industry use modern and faster spinning machinery. These high speed machines require much cleaner raw material, and therefore they constantly challenge the testing machinery developers to provide faster and more accurate testers and quality controllers.

4.1.1   Fibre testing

Latest trend in the fibre testing was the development of a single instrument in which all the major testing parameters can be tested. The instruments we discuss in this category are:

- High volume instrument (HVI).
- Advanced fibre information system (AFIS).
- Fibre contamination tester (FCT).

High volume instrument (HVI)

HVI was developed for the United States Department of Agriculture (USDA) in 1969 [1, 2]. It was designed to be used as a marketing tool to evaluate the quality of the fibre within a bale of cotton. HVI evaluates multiple fibre characteristics in a high volume of samples at a relatively high rate of speed in comparison to hand classing.

First HVI model was commercially manufactured and marketed by Motion Control Inc., USA in 1979. Spinlab, USA developed its HVI (HVI 900) in 1985. Both the instruments give the same parameters of cotton fibres in terms of length, length variability, micronaire, bundle strength, yellowness and greyness.

HVI can be calibrated through two different modes viz. ICC (International Calibration Cotton) mode and HVI mode. Calibration through ICC mode gives 2.5% span length, 50% span length, uniformity ratio(UR%), while calibration through HVI mode gives upper half mean length(UHML), mean length(ML) and uniformity index(UI%).

High volume instrument systems are based on the fibre bundle strength testing, i.e., many fibres are checked at the same time and their average values determined. Traditional testing using micronaire, pressley, stelometre and fibrograph are designed to determine average value for a large number of fibres, the so called fibre bundle tests. In HVI, the bundle testing method is automated. Here, the time for testing is less and so the number of samples that could be processed is increased, quite considerably. The influence of operator is also reduced.

Fibre parameters measured in HVI 900 are as follows:

- Length (staple length/span length), length uniformity (length uniformity index) and short fibre content.
- Strength and elongation.
- Fineness in micronaire.
- Colour (Rd and +b).

HVI 900 consists of following modules joined one after another in series:

- Module 910:Length, strength and maturity.
- Module 920:Micronaire.
- Module 930:Colour.

- Module 935:Trash.
- Module 940:Software for CSP calculation and memory.

HVI –Working principle

The fibrogram method is preferred while preparing the sample for fibre length estimation. The sample has to be presented to the measuring zone by clamping the fibres at a random catch point. Here the fibro sampler is used. The test specimen obtained using the fibro sampler/comb combination is a beard of fibres with individual fibres projecting to different length from the clamping point.

In HVI, strength testing is also done on the same beard of fibres prepared for length measurement. For micronaire testing, a sample of cotton weighing approximately 10 g is used. For colour testing, random mass of fibres sufficient to cover the test window is used for measurement.

Shortcomings of HVI 900

Following shortcomings are found in HVI 900:

1. Short fibres shown are more than 4 mm in length i.e., cannot read fibre length of less than 4 mm.
2. Trash parameters shown by Spinlab are quite different from that of trash% given by Shirley Trash Analyser. HVI testing is based on optical method to identify trash which cannot count below 0.1 inch whereas Shirley Trash Analyser or similar instruments work on the principle of gravimetric measurement of trash, dust and microdust.
3. HVI 900 works on CRE principle whereas Stelometer works on CRL principle for strength measurement.
4. Rate of bundle extension in HVI is faster than conventional tester.
5. In Stelometer fibre crimp and short fibres are removed while preparing samples and most of the cotton fibres take part in loading whereas all fibres do not take part in HVI.
6. While measuring tenacity in g/tex, the mass of the fibres is measured indirectly (optically) in HVI 900 where as it is measured directly in conventional instruments.

Advanced fibre information system (AFIS)

The development of AFIS was the result of cooperative efforts between the USDA Agricultural Research Service at Clemson, SC and Schaffner Technologies, with research beginning in 1982 [3]. One of the primary objectives in the early design of this instrument was the ability to measure trash and neps. This was followed by efforts to measure fibre dimension, number of short fibres and eventually a complete fibre length distribution [3–5]. These properties were chosen because of their value in the fibre-to-yarn engineering process.

Advanced fibre information system is based on the single fibre testing. There are two modules here, one for testing the number of neps and the size of neps, while the other one is used for testing length and diameter of fibre. Both modules can be applied separately or together. With the introduction of AFIS, it is possible to determine the average properties for a sample and also the variation from the fibre to fibre. The information content in the AFIS is more. The spinning mill is dependent on the AFIS testing method, to achieve the optimum conditions with the available raw material and processing machinery. The testing time per sample is 3 min in AFIS-N module. This system is quick, purpose oriented and reproducible.

AFIS –working principle

A fibre sample of approximately 500 mg is inserted between the feed roller and the feed plate of the AFIS-N instrument. Opening rollers open the fibre assembly and separate off the fibres, neps, trash and dust. The trash particles and dust are suctioned off to extraction. On their way through the transportation and acceleration channels, the fibres and neps pass through the optical sensor, which determines the number and size of the neps. The corresponding impulses are converted into electrical signals, which are then transmitted to a microcomputer for evaluation purposes. According to these analyses, a distinction is made between the single fibres and the neps.

Basic difference between HVI and AFIS is that HVI is primarily required for bale management and mixing control but AFIS is needed for process control and has immense practical application to set right the process in right direction from Blow room to Drawing. Following information are obtained through AFIS:

1. Ln/Lw –Mean length of fibres calculated as number/weight basis.
2. SFCn/SFCw – Short fibre content (<12.7 mm) on number and weight basis.
3. UQLW (Upper Quartile Length) – Fibre length in mm exceeded by 25% of the fibres by weight in the test specimen.
4. 5% mm/2.5% mm – Fibre length exceeded by 5%, 2.5% of fibres by number.
5. IFC – Percentage of fibres with less than 0.25circularity. Lower the IFC, better is dyeability.
6. Maturity ratio – The ratio of number fibres with 0.5 (or more) circularity to the number of fibres with 0.25(or less) circularity.

7. Nep – (cnt/g) – Nep count per gram.
8. SCN (cnt/g) – Seed coat count per gram.

Fibre contamination system (FCT)

The fibre contamination tester (FCT) was first described by Mor [6]. More recently, Lintronics [7] proposed a standardised procedure for this equipment. Mor claims that this instrument is able to detect stickiness from any origin (plant sugars, insect honeydew, oily substances, etc.).

The general principle is to feed a fibre sliver, whose mass and length is fixed, into a micro card. The web formed passes between two drums under pressure. The sticky spots adhere to the drum where they are counted.

The FCT system is used for testing stickiness, neps, trash and seed coat fragments. The sample in the form of bundle is fed into a self-cleaning, micro carding device integrated in the FCT, to produce about 10 m of transparent web in order to expose the impurities and contaminants in the best way possible. An area of 1 m2 per sample is tested.

Different methods of maturity measurement

- Double compression airflow measurement.
- Polarised light analysis.
- Causticaire method.
- Centrifugal methods.
- Near infrared spectrometry.
- Image analysis.

4.1.2   Sliver, roving and yarn testing

1. Wrap block.
2. Wrap reel.
3. Electronic count system.
4. Electronic twist tester.
5. Strength tester (Uster Tensorapid, Tensojet, etc.).
6. Evenness tester (Uster Tester model 2/3/4/5, etc.).
7. Classimat tester (Uster Classimat models).

4.1.3   Other testing equipment

- Electronic balance.
- Trash analyser.

- Maturity tester.
- Yarn's appearance tester (for Black Board).
- Stroboscope.
- Tachometer (contact type).
- Electronic moisture meter.
- Pneumafil suction pressure (ASPI-ATIRA suction pressure instrument).
- Nilometer for Drawing top roller pressure.
- Tarp gauge (top arm pressure gauge for simplex and ring frame).

4.1.4   Fabric inspection

Fabric is inspected to determine its acceptability from a quality view point. There are various fabric inspection systems such as 4 point system, 10 point system, 2.5 point system, etc. 4 Point system, a standard established under **ASTM D5430 – 07(2011),**is the most popular and broadly used system due to its practical, impartial, and worldwide acceptance. Most of the apparel industries prefer 4 point rating system for determining fabric quality, and it is certified by the American Society for Quality Control (ASQC) as well as the American Apparel Manufacturers (AAMA).

The 4 point system (Table 4.1) assigns 1, 2, 3 and 4 penalty points according to the size, quality and significance of the defect. A defect can be measured either in length or in width direction. Length of piece taken for inspection is 100–120 yd. Whenever errors are recognised during fabric inspection under 4points system, the defect must be assigned a number of points depending on the severity or length as shown in Table 4.1:

Table 4.1: 4 Point system

Fabric defects in inches (")	Points
From 0"> 3" length/width	1 point
From 3.1" > 6"length/width	2 points
From 6.1" > 9" length/width	3 points
More than 9" length/width	4 points

Fabric is then graded as follows:
- First grade quality – Less than 4 points/10 yd or 40 points/100 sq. yd.
- Second grade quality – Less than 4 points/5 yd or 80 points/100 sq. yd.
- Unmerchantable grade – More than 80 points/100 sq. yd.

4.1.5   Standard methods of fabric testing

1. Bending length (IS 6490 1971): Bending length equals half the length of rectangular strip of fabric that will bend under its own weight to an angle of 41.5°. It is also equal to the length of a rectangular strip of materials that will bend under its own weight to an angle of 7.1°. It is expressed in centimetres.

2. Wettability of cotton fabrics (IS 2349 (1963):

    A drop of water is allowed to fall on the test specimen. A lamp is suitably placed so that the image of the lamp is seen on the drop. The time taken for the image of the lamp to just disappear at the edge of the drop is noted. This time is a measure of the wettability of the sample under test. Longer the time taken lower is the wettability.

3. Colour fastness to washing at 60°C and 95 °C (IS 764 1979):

    A specimen of the textile in contact with one or two specified adjacent fabrics is mechanically agitated under specified conditions of time and temperature in a soap, or soap and soda solution, then rinsed and dried. The change in colour of the specimen and the staining of the adjacent fabric, or fabrics, are assessed with reference to the original fabric, either with the grey scales or instrumentally.

4. Colour fastness of textile materials to rubbing (IS 766 1988): Specimens of the textile are rubbed with dry and wet rubbing clothes. Two alternative sizes of rubbing finger are specified, one for pile fabrics and the other for other textiles. The staining of the rubbing clothes is assessed with the grey scale.

5. Colour fastness of textile materials to perspiration(IS 971 1983): Specimens of the textile in contact with adjacent fabrics are treated in two different solutions containing histidine, drained and placed between two plates under a specified pressure in a testing device. The specimens and the adjacent fabrics are dried separately. The change in colour of each specimen and the staining of the adjacent fabrics are assessed with the grey scale.

6. Colour fastness of textile materials to weathering by xenon arc lamp (IS 6152 1985):

    Specimens of textile are exposed under specified conditions to light from a xenon arc lamp and to water spray. At the same time and in the same cabinet, eight standard patterns of blue wool cloth are exposed to light but are protected from water spray by a sheet of window glass. The fastness is assessed by comparing the change in colour of the textile with that of the standard patterns.

7. Whiteness – based on Hunter whiteness scale: It is based on the percentage reflectance of light from the fabric surface. Higher the reflectance, the whiter is the fabric.
8. Fabric pilling –method ASTM – D1375:
   Martindale tester -

Grade	Pilling intensity
5	No pilling
4	Slight pilling
3	Moderate pilling
2	Severe pilling
1	Very severe pilling

4.1.6   Source of fabric defects

Fabric defects come from any one or more of the sources listed below:
1. Spinning faults.
2. Ring doubling faults.
3. Winding faults.
4. Weaving preparatory faults.
5. Weaving faults.
6. Knitting faults.

Types and nature of faults, those are generated in the fabric due to:
1. Spinning: Coarser ends/picks, finer ends/picks, slubs, hairy, snarls, neps, stain/contamination, thick/thin, warps streak, weft bars, etc.
2. Ring doubling: Single yarn, untwist/less twist yarns, snarls, stain, knots, uneven twist setting (all for ply yarns).
3. Winding package: Slough off, bunches, parallel yarns, etc.
4. Weaving preparatory: Multiple breaks, sizing stains, wrong drawing-in, sticking ends, mixed yarn, missing ends, loose selvedge, etc.
5. Weaving: Missing ends, missing picks, warp streaks, reed marks, reediness, crammed picks, lashing-in, weft bars, temple marks, etc.
6. Knitting: Holes, drop stitches, vertical stripes, horizontal stripes, barre, etc.

4.1.7   High seed coats in Indian cotton – its problems and remedies

Trash (mainly seed coats) in raw cotton plays a very important role to decide the thick places and neps in the yarn. ATIRA's work in this field depicts that thick places in cotton yarn are due to:

1. Fibre dust.
2. Fibre dust with foreign matters.
3. Fibre dust with foreign matters and flies.
4. Fibre dust with flies.

Microscopical analysis reveals that fibre dust with foreign matters contribute 60–80% of the total thick places in the yarn. Similarly neps are classified into four types:

1. Solid.
2. Peripheral.
3. Loosely attached.
4. Wrapper fibres.

These neps are further classified for their compositions and classified into three categories namely neps due to:

1. Foreign matters.
2. Immature fibres.
3. Fibre dust.

Foreign matters with immature ovules contribute to 75% of the total neps in the yarn. Old methods of ginning process leave behind a good percentage of seed which include defective seeds too in the raw cotton. Removal of such seeds from the raw cotton is mainly through spinning process at blow-room and carding stage. If the colour of the seed is black then removal is easy but if the colour is brown, its removal becomes difficult. Furthermore, if micronaire is finer with low percentage of maturity (below 75%) then it becomes much more difficult.

Seed coats should be dropped at blow-room stage through (1)Axi flow type beater and (2)Step cleaner type beater through grid bars. Grid bars setting at these two stages are very much important.

In carding fragmented or broken coats which are not removed at blow-room stage are to be removed effectively through closer setting of combing bar segments under licker-in and back stationery flats. Closer setting of front stationery flats helps to control actual neps.

Excess trash/seed coats/kitties in final yarn gives problem after dyeing as seed coats/kitties gives white specks in the yarn due to lower absorption of dyes.

4.1.7.1   How to measure kitties/sq ft in the knitted fabric

For each and every count and lot one knitted fabric is to be made on single jersy weft knitting machine. On tubular face of the fabric, one sq ft area is pen marked and divided into four equal sections. From each section, kitties are to be counted manually by naked eyes. Then number of kitties as found in four sections are to be added up to get kitties/sq ft. This study is to be repeated from another place of the knitted fabric and take the average value.

The biggest draw back in the Indian cotton is the presence of high seed coats and trash along with high level of contamination. Presence of high percentage of short fibre and low elongation of cotton fibres also create problems. Foreign buyers are very much critical about contamination and often show dissatisfaction and even suit for compensation.

4.1.7.2 *Presence of high seed coats (kitties) in cotton yarn of Indian origin (This practice was used by the author, yielding very good response from the customer)*

Excess seed coats (kitties) in Indian cotton give a problem for knitted fabric appearance particularly for carded yarn. Very often, complaints come from the foreign buyers. If following standard for kitties/sq ft in knitted fabric is maintained, then the complaint from the market seldom comes. Standard is given in Table 4.2:

Table 4.2: Seed coats (kitties) in cotton yarn

Count (Ne)	Kitties/sq ft in the knitted fabric	
	Carded cotton	Combed cotton
20s	300–350	25–30
30s	200–250	20–25
40s	100–150	15–20
Above 40s		10–12

4.2 Yarn irregularity

The most important object of a spinning mill is to produce yarns of uniform character such as uniformity in weight per unit length, diameter, turns per inch, strength, etc. It is practically impossible to produce a perfect uniform yarn. Transforming millions of staple fibres into a perfect uniform yarn is extremely hypothetical, more particularly in case of natural fibres of varying fineness, maturity, length, colour, diameter, etc. Variation in weight per unit length is the basic irregularity in yarn. All other irregularities are dependent on it. This is because weight per unit length is proportional to fibre number i.e.; number of fibres in cross section of yarn.

4.2.1   Importance of yarn irregularity

While a yarn may vary in many properties, evenness is the most important quality aspect of a yarn, because variations in other yarn properties are often a direct result of yarn count irregularity (variation in mass per unit length along the yarn). It is well known that twist tends to accumulate at the thin places in a yarn, so irregularity in yarn count will cause variations in twist along the yarn length. This preferential concentration of twist in thin places along a yarn also exacerbates the variations in yarn diameter or thickness, which often adversely affects the appearance of the resultant fabrics.

Irregularity can adversely affect many of the properties of textile materials. The most obvious consequence of yarn unevenness is the variation of strength along the yarn. If two yarns have same average mass per unit length and twist level, but one yarn is less regular than the other, it is clear that the more even yarn will be the stronger of the two. The uneven one should have more thin regions than the even one as a result of irregularity, since the average linear density is the same. Thus, an irregular yarn will tend to break more easily during spinning, winding, weaving, knitting or any other process where stress is developed. However, due to the twist migration, thin place may not be necessarily weak places.

A second quality-related effect of uneven yarn is the presence of visible faults on the surface of fabrics. If a large amount of irregularity is present in the yarn, the variation in fineness can easily be detected in the finished cloth. The problem is particularly serious when a fault (i.e. a thick or thin place) appears at precisely regular intervals along the length of the yarn. In such cases, fabric construction geometry ensures that the faults will be located in a pattern that is very clearly apparent to the eye, and defects such as streaks, stripes, barre or other visual groupings develop in the cloth. Such defects are usually compounded when the fabric is dyed or finished, as a result of the twist variation accompanying them.

Twist tends to be higher at the thin places in a yarn. Thus, at such locations, the penetration of a dye or finish is likely to be lower than at the thick regions of lower twist. In consequence, the thicker yarn region will tend to be deeper in shade than the thinner ones and, if a visual fault appears in a pattern on the fabric, the pattern will tend to be emphasised by the presence of colour or by some variation in a visible property, such as crease-resistance controlled by a finish.

Other fabric properties, such as abrasion or pill-resistance, soil retention, drape, absorbency, reflectance, or lustre, may also be directly influenced by yarn evenness. Thus, the effects of irregularity are widespread throughout all areas of the production and use of textiles, and the topic is an important one in any areas of the industry [8].

4.2.2   Components of yarn irregularity

Yarn irregularity has been classified as totally random variations, periodic variations and quashiperiodic variations.

4.2.2.1   *Totally random variations*

It is known that spun yarns have a certain minimum irregularity. Martindale [9] viewed this as arising out of a totally random arrangement of fibres, i.e. the fibre left-ends forming a Poisson process. Variations arising out of such an arrangement are referred to as totally random variations. These depend only on the fibre characteristics and the mean linear density. Purely random irregularity forms an unavoidable component of total irregularity, so that a minimum achievable random irregularity can be acceptable for apparel usage.

Random variation is the variation which occurs randomly in the textile material, without any definite order. Suppose a yarn is cut into short equal lengths, say, of 1 inch, and weight of each consecutive lengths are found out. The weights are plotted in a graph against the lengths similar to Fig. 4.1 shown below:

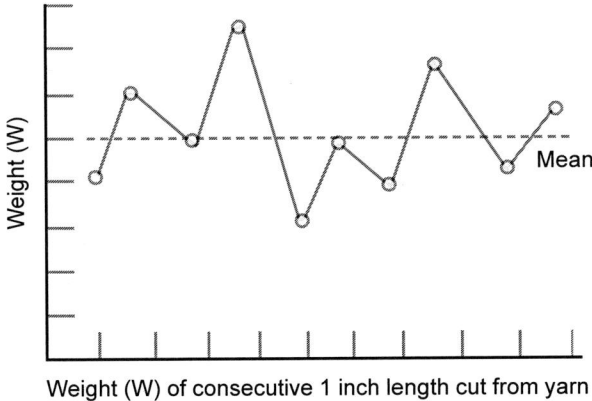

Weight (W) of consecutive 1 inch length cut from yarn

Figure 4.1: Random variations in yarn

4.2.2.2   *Periodic variations*

Periodic yarn faults, as shown in Fig. 4.2, are thick and thin places, which always occur with the same distance from each other. Such faults are caused in the spinning process, when yarn guiding elements are defective.

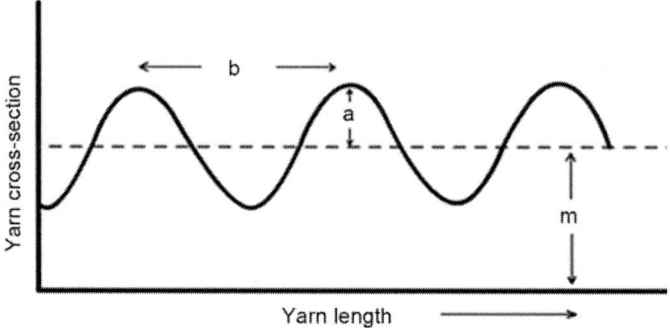

Figure 4.2: Periodic variations in yarn

Wave length and amplitude:

m = mean chart height.

a = amplitude of wave or variation.

b = wave length periodic variation.

$(a/m) \times 100$ = percentage amplitude.

Variations produced by factors such as eccentric rollers or broken gears create a regular pattern of waves along the length of the yarn. These may be represented as:

$$A_j cos W_j x + B_j sin W_j x, j = 1, \ldots, p,$$

where $\sqrt{(A_j^2 + B_j^2)}$ and W_j are respectively the amplitude and angular frequency of the jth period.

Periodic mass variations in yarn can cause weft bars, diamond barring effects, moiré effects, and weft stripes or rings in the resulting fabric. Hence, periodic irregularity should not be permitted at all, since it greatly affects the appearance of fabric and must be controlled.

Depending upon the wavelength of the periodic fault, the mass variations are classified as:

Short-term variation (wavelength ranges from 1 to 50 cm).

Medium-term variation (wavelength ranges from 50 cm to 5 m).

Long-term variation (wavelength longer than 5 m).

Periodic variations in the range of 1–50 cm are normally repeated a number of times within the woven or knitted fabric width. This results in the thick or thin places which will lie near to each other. This produces, in most cases, a "Moire Effect". This effect is particularly intensive for the naked eyes if the finished product is observed at a distance of approx. 50 cm to 1 m.

Periodic mass variations in the range of 50 cm to 5 m are not recognisable in every case. Faults in this range are particularly effective if the single or double weave width or the length of the stretched out circumference of the knitted fabric, is an integral number of wave-lengths of the periodic fault, or is near to an integral number of wave-lengths. In such cases, it is to be expected that weft stripes will appear in the woven fabric or rings in the knitted fabric.

Periodic mass variations with wave-lengths longer than 5 m can result in quite distinct cross-stripes in woven and knitted fabrics, because the wave-length of the periodic fault will be longer than the width of the woven fabric or the circumference of the knitted fabric. The longer the wavelength, the wider will be the width of the cross-stripes. Such faults are quite easily recognisable in the finished product, particularly when this is observed from distances further away than 1 m.

4.2.2.3   *Quasiperiodic variations*

All variations which are neither periodic nor totally random have been termed quasiperiodic. Generally, a large part of this component is due to drafting wave. However, drafting is known to introduce variations other than drafting waves, the nature and cause of which are not clearly understood today. Even the description of the drafting wave as given by Foster [10]does not seem to lend itself to a strict mathematical definition.

The basis cause for the formation of drafting waves is "floating fibres". Floating fibres are fibres which are not gripped by any roller pairs in the drafting zone. Also, improper roller setting, improper roller pressure, high drafts, etc. increase the amplitude of drafting wave.

How to calculate drafting wave

Wave length of "Drafting Wave" at which the maximum amplitude occurs is related with "Mean Staple Length" of constituent fibre as follows:

$\lambda m = KL^*$

where,

λm = Average wave length (highest amplitude),

L^* = Average staple length,

K = Constant.

The value of K of different processing stages is given below.

Processing stage	K
Yarn	2.75
Roving	3.5
Drawing	4.0

The value of wave length changes with the subsequent and progressive drafts in the spinning process. For example, drafting wave appeared in the roving of a simplex machine will be 3.5 ×average staple length. This roving end after being drafted through ring frame with a draft of 30 will be attenuated 30 times and appear in this spectrogram as drafting wave to that extended wavelength.

In this connection, one write up by SITRA [11], is quoted here:

The average wave length serves as a base to locate the source of drafting waves. If the average wave length of a sliver is equal to K times, the average staple length (the value of K as given above) then it indicates that the drafting waves have been introduced in the latest drafting zone of the processed material. For example, if the average staple length of a fibre material is 38 mm and when the draw frames sliver made from these fibres is tested for spectrogram, the average wave length of the drafting wave is 15 cm. This means that drafting waves have been introduced in the front zone of the draw frame since the average wave length is equal to 4 mm× 38 mm (K× average staple length: K value for sliver = 4).

If the drafting waves have been formed in the back zone then the average wave length would have been higher by the following draft. For instance, in a two zone draw frame, drafting waves are formed at the back zone for the above sliver. If the front zone draft is 6, then the drafting waves are indicated in the spectrogram of the draw frame sliver as six times the drafting wave at a point of formation. Having identified the zone, the direct cause responsible for the drafting waves can be identified e.g. improper roller setting, high break draft, etc.

4.2.3   Causes of irregularity

Mass variation can be attributed to the properties of raw materials, inherent short comings in yarn making and preparatory machines, mechanically defective machinery and/or external causes as a result of working conditions and improper housekeeping [12]. Purely random irregularity forms an unavoidable component of total irregularity, so that a minimum achievable random irregularity can be acceptable for apparel usage. The periodic irregularities which are found in the spun yarns may be the result of machinery defects such as eccentric drafting rollers, variability in the covering of drafting rollers, inaccurately cut or worn out drafting rollers and the vibration of drafting rollers. Yarns which are affected by any of these defects occurring in the drafting prior to spinning can appreciably affect the yarn and the resulting fabric [13]. Following are considered to be causes of yarn mass variation:

1. Properties of raw material,
2. Fibre arrangement in the yarn,
3. Fibre behaviour,
4. Inherent shortcoming of machinery,
5. Mechanically defective machinery,
6. External factors such as working condition and inefficient operation.

Cotton-fibre characteristics influence the evenness of yarns spun from these fibres. Cotton fibres are not of same length and are not arranged straight and parallel in the strand. Lunenschloss and Helli [14] show major variations in the frequency of thick places as a result of using different types of cotton and also between different monthly consignments of supposedly identical cotton spun into yarns of constant count. Their work suggests that spinning conditions may be more important than raw-material characteristics and that care should be exercised in drawing conclusions regarding the influence of fibre characteristics on evenness.

Czaplicki [15] examines the effect of the maturity of cotton fibres on irregularity and concludes that processing conditions also exert some influence. He finds that mature fibres give fewer faults than immature ones at the carded-sliver stage and after spinning on a ring frame but that yarns spun on an open-end-spinning system are not influenced by the degree of maturity of the fibres. Seshan et al. [16] support the latter conclusion.

An expression, as shown below, has been derived by Ratnam et al. [17] for yarn irregularity in terms of fibre properties and processing parameters. The agreement between the actual and predicted irregularity was found to be very close, the correlation coefficient being 0.98.

$$V^2 = \{29.4(F/L)^2 + 3.62\}\,(D-1) + V_r^2$$

where,

V^2 **and** V_r^2: Relative variance of yarn and roving irregularity, respectively,

D: Ring frame draft,

L: 50% span length,

F: Fineness (μg/in.)/ maturity coefficient.

C: Roving hank.

A part of the added variance at the ring frame can be attributed to the machine condition or the type of drafting in ring-spinning. The square of the fineness/maturity coefficient (F) divided by the square of the 50% span length (L) is highly correlated with the contribution of fibre properties to yarn irregularity ($r = 0-97$). In view of this very close agreement, the factor F/L, which would approximately depend on the ratio of fibre diameter to fibre length, can be taken as a measure of cotton quality for yarn irregularity.

The added variance in ring frame drafting is independent of the level of input irregularity. The index of irregularity explains only a small part of the variation in yarn irregularity and as such is not a suitable measure for judging spinning performance. It is necessary to take into account both the fibre length and the length uniformity.

Textile fibres are not rigid. Their manipulation during conversion into yarn is an immensely complex combination of mechanical movement which usually requires some degree of compromise. The desirable results of relocating large number of fibres at high speed and arranging in well-ordered form tend to be difficult. Fibres assembled into the form of a twisted strand constitute a yarn. Fibres are not precisely laid end to end, and gaps are present between them. As a result of yarns twist, fibres arrange in spiral form in a series of folds, kinks and doublings.

The regularity of a yarn fundamentally depends on fibres and their arrangement within the yarn [18]. The fibres constituting the yarn are arranged in a completely random way during blending, carding, doubling, roving and spinning processes. Therefore all the fibres have the same chance of being found at any selected place in the yarn. Therefore, the fibre length variation causes irregularity in yarn cross-section.

Given a random fibre distribution of fibre leading ends, the best that any roller drafting machinery can do is to preserve the original random fibre arrangement up to and including the spinning stage. Such an arrangement would involve 'Perfect' drafting in which all fibres including floating fibres would move at the back roller speed until the leading end of each fibre was gripped by the nip of the front rollers and instantaneously accelerated to the front roller surface speed [19].

Unfortunately 'perfect' drafting is unattainable because of the lack of positive control of the floating fibres. These are acted on by frictional forces due to contact with other fibres which are moving at speeds between those of the back rollers and the front rollers; it is on these frictional forces that the motion of each floating fibre depends.

In drafting process the short fibres move in groups causing non-random wave-like patterns called drafting waves, which are responsible for periodic thin and thick places over a yarn length [20]. In practice some fibres released from the back rollers tend to accelerate before they reach the front roller nip; short fibres tend to do so in bunch, forming a thick place. When such a thick place arrives at the front rollers a greater force is necessary to draft it and so the following portion between the rollers is drafted to a greater extent thereby forming a thin place. When the thin place later reaches the front rollers, the drafting force decreases and so the thin place will be followed by another thick place and so on.

Fibre shape directly affects yarn regularity. The fibre cross section, arrangement of fibre section and space between the fibres will vary from yarn section to section. Hence the mass of each section will differ. A thin place in yarn will have lower mass and less strength. In thin regions, yarn twist tends to be higher since resistance to deformation is lower.

Fundamentally spun yarn production involves twisting of a random fibre array. Twisting tends to condense the yarn structure into an irregular close-packed polygonal shape, but the cross-section still possesses a concave–convex irregular shape [21]. In addition, there is a complex relationship between yarn diameter, twist and mass. Therefore, it is hard to predict the effect of twist on yarn structure. For example, twist is not constant but accumulated in the thinnest parts of the yarn. Therefore, high twist compresses thin places and exaggerates the variations in the apparent diameter [22].

Neps are caused by foreign elements, immature fibres and insufficient and improper cleaning during preparation processes. These faults are usually random and visible to the human eye. They are detected by many evenness-testing instruments. When a cross section deviation exceeds a preset value, the instrument classifies the imperfection as either a nep, or a thin or thick place. The standard levels are +200%,−50% and +50%, respectively. The length of the fault is usually in the order of a few centimetres [23].

Since machines even in good condition produce irregular yarns, it is reasonable to assume that defective machinery will increase the amount of irregularity. An eccentric front roller of the ring spinning machine leads to a periodic fault with a wavelength of 8 cm, as this roller always causes faulty drafts in the draw-box within the same time intervals. The size of each individual fault is mostly not disturbing. But as a series of yarn faults, they can very well be disturbing. In most cases, disturbing periodic faults are formed at the ring-spinning machine. Widely known are defects caused by cuts and pressure marks on the take-off cylinder. By this, the continuous distribution of the fibres is disturbed, which results in thin and thick places. The size of the fault corresponds to an alteration/shift of all fibres of about 30–50%. The fault length depends on the dimension of the defective machine part. The distance between the single events corresponds to the circumference of the roller, e.g. at the front roller of a draw-box. A further reason for periodic faults can be pressure marks on the top roller. If a spinning position or the whole spinning frame is stopped and the pressure is not taken from the top roller, it can lead to pressure marks on the top rollers after longer stops and thus to periodic defects in the yarn(seldom occurring).

The implementation of an efficient maintenance system is essential if the level of irregularity is to be kept within bounds. Machines drift out of adjustment, bearings become worn, components get damaged, and lubrication systems clog and dirt works its way into the mechanism. Faulty rollers (top roller eccentricity) and gear wheels usually produce periodic variation.

Pillay and Hariharan [24] worked exhaustively on causes of yarn faults and concluded:

1. The amount of trash present in the cotton influences significantly the formation of faults during spinning.

2. Man-made fibres produce a lesser number of faults than cotton fibres when spun under identical conditions. The surface nature of fibre has an appreciable effect on fault formation.

3. A higher degree of opening in the blowroom and good cleanliness and maintenance of machinery reduce the fault formation.

4. Better fibre individualisation in carding, as carried out during tandem carding, reduces the incidence of faults appreciably.

5. The number of faults in cotton yarn decreases with increase in the percentage of noil removed during combing. The reduction in total faults for every 1% level of waste removal is 4.4%.

6. The use of overhead cleaners during spinning reduces the formation of faults.

7. The fibre control in spinning is of extreme importance not only in terms of irregularity, but also in terms of number of disturbing faults in yarn.

4.2.4   Effects of irregularity

The topic of yarn evenness, though of some interest in itself, would be of minor importance if it were not to the significant effects that the presence of yarn irregularity produces. It has long been recognised that evenness is the fundamental cause of a number of defects in both yarn and fabric and that these defects are of great concern because they can render an otherwise perfectly satisfactory fabric totally unacceptable. Irregularity affects strength of yarn due to the presence of more thin places. Thin places in sliver, roving or in yarn will be weak places and the greater will be the chance of breakage for more irregular yarn.

Processability of the material is affected by yarn irregularity. For example passing thick places through heald eye or reed in weaving, needle eye during knitting or in sewing machine leads to yarn breakage.

Yarns free from strong periodic variations but with a high degree of general irregularity will tend to produce patchy fabric. Under certain conditions yarns with periodic thick and thin portions will cause the fabric to exhibit an unwanted pattern. In warp it gives streaky appearance and in weft gives diamond bars and block bars.

One effect of yarn irregularity on the dyeing process is the thicker and the softer parts of the yarn take up more size than the thinner and harder region; after the desizing process prior to dyeing, the distribution of the residual size may be uneven and cause difficulty in achieving a level dyeing.

Unwin and Reast [25] in a paper dealing with the effect of yarn irregularities on fine-gauge fashioned hosiery, summarise the many potential causes of unsatisfactory yarn (including those related to evenness) and give details of the effects likely to be produced. In the latter category, they include visual faults in the yarn or fabric structure, irregularity of dyeing, barre stripes and weak spots, all of which can be assumed to owe at least part of their presence to the incidence of irregularities in the yarn. Since the review is a very general one, however, there is little location of specific causes to any one fault.

McFarlane [26], in a consideration of the effect of yarn irregularities on viscose fabrics, points out that different types of irregularity, even those caused at different stages in the production process, can give rise to similar faults in the finished fabric.

Bowles [27] discusses the effects of yarn unevenness on the quality of woven or knitted fabrics. He includes barriness as the major problem in the former materials and suggests that irregularity is more crucial in weft-knitted fabrics. Piso [28] identifies yarn-evenness defects as the major cause of fabric problems and includes in his list barre in knitted goods, weft bands in woven fabrics, streaks in carpeting, hosiery runs and poor breaking strength in tyre yarns.

Snowden and Sidi [29] examine cloths produced from a range of yarns blended in a variety of ways and find that short-term yarn irregularity gives short weft streaks in the cloth.

Bleakley [30] discusses the importance of regularity in linen yarns and makes the somewhat unusual statement that diameter variation is a desirable property, since it maintains the truly Linen' characteristic appearance of the cloth. He then points out, however, that this desirability has limits, since cloth with serious yarn unevenness incorporated in it is liable to suffer from disastrous defects, such as stripiness.

Zellweger Ltd. [31] devotes an Uster News Bulletin to the relationship between irregularity and the appearance of finished fabrics. In it, a direct correlation between the value of $U\%$ and the visibility of unevenness defects in a knitted fabric is demonstrated.

Mahajan [32] investigates three different causes of periodic count variation responsible for weft bars and notes the relationship between the two factors. Garde et al. [33] find moderate agreement between yarn-unevenness values and the subjective rating of the appearance of fabric woven from the yarns.

Perhaps the most important effects of yarn unevenness, however, arise when dyeing is carried out. The presence of colour can emphasise visible defects, and a fabric that is acceptable in the grey state can become totally undesirable once dyeing has taken place, purely as a result of yarn-irregularity faults.

Weight irregularity is identified as the major source of defective appearance in dyed or finished cloths, but other factors (including variations in twist, fibre length, cotton maturity, or natural shade) can complicate the situation. In addition, the open spaces arising in the cloth as a result of variation in yarn thickness can also influence dye penetration (and hence colour) uniformity, apart from the fact that they present a visually uneven appearance to an observer.

Warty and Talele [34] analyse problems in wet processing arising from yarn irregularities by dividing them into warp wise or weft wise defects and then classifying the defects as bars, streaks, or lines.

4.2.5   Methods of measuring yarn evenness

The estimation of yarn evenness has long occupied the attention of textile personnel, mainly because of the importance of the topic with reference to the effects, whether visual or influencing mechanical properties, on the fabric produced from the yarn. In fact, to measure irregularity, many methods are available involving from no equipment to electric instruments.

4.2.5.1   Subjective method

Yarn appearance board (visual examination):
The estimation of yarn evenness has long occupied the attention of textile personnel, mainly because of the importance of the topic with reference to

the effects, whether visual or influencing mechanical properties, on the fabric produced from the yarn. Before the advent of methods of measuring evenness quantitatively, subjective estimates were commonly made by winding yarns on a black card and looking for visible faults.

Yarn to be examined is wrapped onto a matt black surface in equally spaced turns as shown in Fig. 4.3. The black boards are then examined under good lightening conditions using uniform non-directional light. ASTM has a series of Cotton Yarn Appearance Standards which are photographs of different counts with the appearance classified in four grades. The test yarn is then wound on a blackboard approximately 9.5 × 5.5inches with the correct spacing and compared directly with the corresponding standard. Motorised wrapping machines are available: the yarn is made to traverse steadily along the board as it is rotated, thus giving a more even spacing. It is preferable to use tapered boards for wrapping the yarn if periodic faults are likely to be present. This is because the yarn may have a repeating fault of a similar spacing to that of one wrap of yarn. By chance it may be hidden behind the board on every turn with a parallel sided board whereas with a tapered board it will at some point appear on the face.

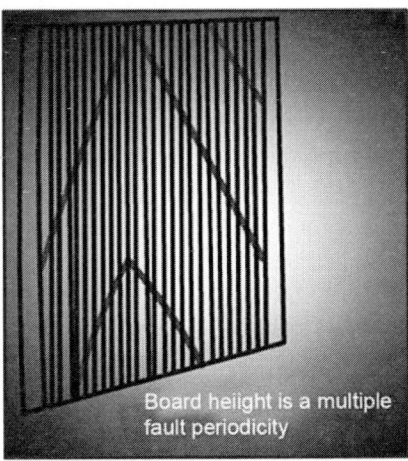

Figure 4.3: Visual examination of yarn wrapped on a black board

Subjective measuring technique provides important additional information that can be correlated with the appearance to be expected in fabrics made from yarn. Grading after viewing a sample of yarn wound with a designated traverse (depend on count) on a black board. ASTM standard test method describes the yarn appearance into five grades. The board is compared with standard photographs and then graded.

Grade A: No large neps, very few small neps, must have very good uniformity, less fuzziness.

Grade B: No large neps, few small neps, less than three small pieces of foreign matters per board, slightly more irregular and fuzzy than A.

Grade C: Some large neps and smaller neps, fuzziness, foreign matters more than B, more rough appearance than B.

Grade D: Some slubs (more than three times diameter of yarn), more neps, larger size neps, fuzziness, thick and thin places, foreign matters than Grade C yarn, overall rougher appearance than C.

Grade E: Below grade D, more defects and overall rougher appearance than grade D yarn.

Louis [35] reports work intended to compare three methods of visual assessment by using results obtained by the Uster technique to check the accuracy of each one. He uses visual assessment of yarns wound on a blackboard, with the aid of ASTM panel photographs, as the basis for his first subjective method. He uses an order of ranking for the second one and a series of paired comparisons for the third. He reports that the first method is not sensitive to small differences in yam appearance and that the Uster data are adequate for indicating grades provided that a narrow range of yarn counts is involved in each evaluation.

4.2.5.2   Mechanical methods

(A) Gravimetric (cutting and weighing) method:

The first method of quantitative measurement, in chronological terms, is the technique of cutting and weighing. The plot of mass against length (cm) is shown in Fig. 4.4. The origins of the technique appear to be lost in history, but it first appears in the literature, as far as can be ascertained, when Barker [36] predicts its usefulness in a paper discussing the theoretical basis of mass-per-unit-length measurements in studying yarn quality. Later, Martindale [37] made use of cutting and weighing elements of yarn in his description of a 'new' method of measuring yarn irregularity. Since that time, the cutting-and weighing technique has been regarded as the basic one against which all others should be calibrated. It does, however, have certain drawbacks, apart from the enormous amount of time that it requires. The precise cutting of a fixed small length (usually 1 cm is regarded as the minimum feasible size) is complicated by the fact that yarns are subject to stretching and requires some care in the exact placement of the cutting device, which must also be well-maintained to ensure that it is consistently sharp and accurate. Nevertheless, the cutting-and-weighing method is still used to check the results of methods that are much more sophisticated in nature.

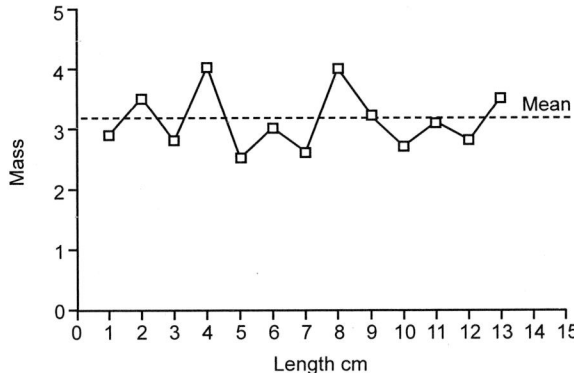

Figure 4.4: Mass vs. length plot in cutting and weighing method

(B) By measuring variation in thickness under compression:

The other mechanical methods of detecting yarn evenness involve devices that respond to changes in yarn diameter, usually by means of feelers that move in response to such changes, the movement being subsequently amplified by suitable means to give a visible trace, which is proportional to the instantaneous yarn diameter. The best-known of these is the Pacific Tester [38] though the earliest instruments derived appear to be the ones reported by Frenzel [39] and Oxley [40] and many others have subsequently been reported. In general, all of them use a light lever, resting on the yarn (often contained within a groove to prevent errors because of spreading), which is forced upwards as thick places are sensed by it and moved downwards again as thin places are presented to it. The motion is transmitted by a series of levers, or by the piston of a micrometre dial, or by some electromagnetic or electronic amplification system, to a device on which the yarn profile may be observed visually or may be obtained as a permanent chart record.

Thickness-gauge type evenness tester

(a) The fixed shoe type: A sample is supplied on or under a fixed shoe and a proper weight is placed on or under the sample. Then the displacement of the weight caused by yarn thickness irregularity is measured after it is magnified by means of a dial gauge [51] or a mechanical [41] or optical lever [42, 53].

(b) The grooved roller type: The fixed shoe type mentioned above can be used for high twist yarns, but it is not to be used for slivers and rovings because the object to be measured becomes obscure as the sample is drawn, broken or flattened by the friction force. The grooved roller type has been devised to remove these defects. Its principle is as follows: a sample is placed in a groove of the roller, the grooved roller is rotated positively, and the displacement of the weight roller by the sample is measured after it is amplified by a mechanical [42] or electric apparatus [46, 52]. Both Saco Lowell tester [54] and Pacific tester [47, 52] are practical applications of this principle.

4.2.5.3   *Air-micrometer (pneumatic) type*

As illustrated in Fig. 4.5, when a sample passes through the outlet nozzle N2 and air of constant pressure is supplied from the inlet nozzle N1, the pressure difference between N1 and N2 is varied by the variation of outlet resistance caused by the passing sample. Such variation of the pressure difference is converted into a height of water or mercury column or into a variation of electric current with an electric resistance strain gauge [49]. There are two types of outlet nozzles: one-side nozzle and both-sides nozzle, which are illustrated in Fig. 4.6. To keep the errors resulting from air-leakage through the yarn-guide nozzle minimum, various means are employed, such as adjusting the diameter of the nozzle to the yarn counts [55], making a labyrinth[59], or immersing the nozzle in a mercury bath [49] so far as the one-side type nozzle is concerned. A short-term variation can be measured with a one-side type nozzle. When a both-sides nozzle is used, the period of variation detected becomes longer, though the errors due to air leakage can be avoided [43]. As to the recording apparatus, a bellows and a pen [49, 56], a bellow and an inductance, or a combination of an electric resistance strain gauge and ink-writing oscillograph can be used. The electric resistance strain gauge has a great practical value from the point of view of dynamic character.

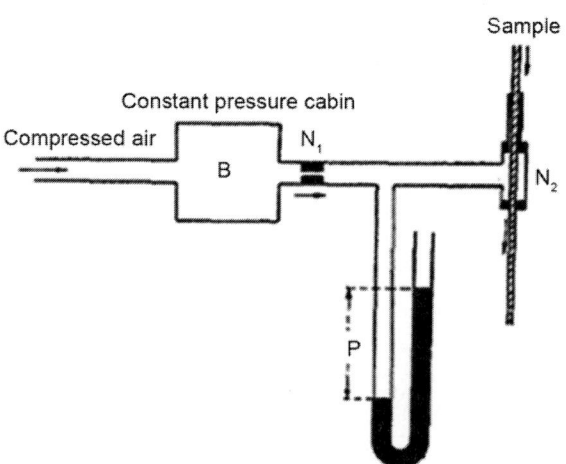

Figure 4.5: Principle of air-micrometer type evenness tester

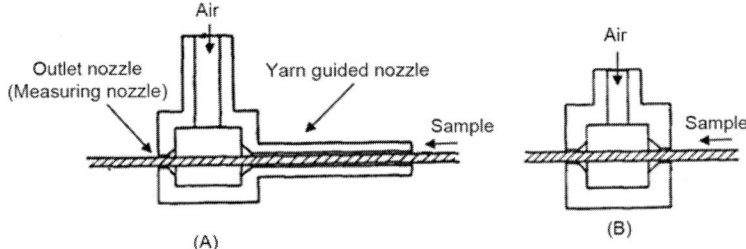

Figure 4.6: Type of nozzles. (A) One side nozzle and (B) Both sides nozzle

4.2.5.4   *Electronic capacitance type tester [61]*

Unevenness is detected from variation of electric capacitance generated by a sample which passes through a gap of the fixed air condenser (Fig. 4.7). The variation of capacitance, which is proportional to the variation of the weight per unit length of a sample under proper conditions, is amplified by the resonant circuit method or by the AC capacity bridge method [58]. Practical applications of this type are Uster [50] and Fielden-Walker [44, 45, 60]. Defects of the capacitance type are that the moisture content of yarns affects greatly the values observed, because the dielectric constant of water is ten times as great as that of fibres. Furthermore, a different mixing ratio of fibres gives a different result of measurement if each fibre has a different dielectric constant. The frequency of oscillator also varies with the temperature. Nevertheless, this type has been used universally, because it has a high sensitivity, is easy to use and can be easily fitted with the calculating apparatus. Recently, it has been used for measurement of lap evenness as well.

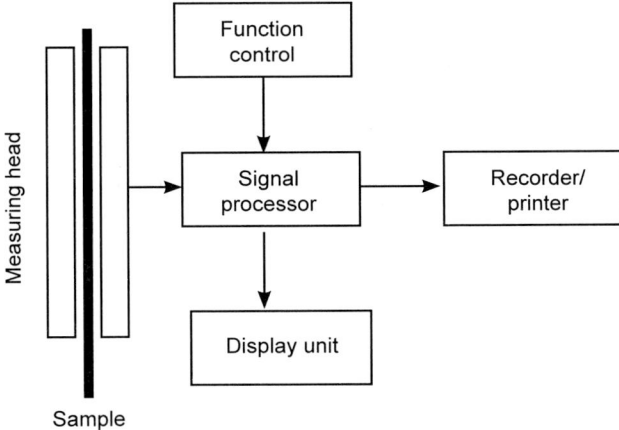

Figure 4.7: Principle of electronic capacitance type tester

In case of Uster Evenness Tester (UT-3) materials under testing in the form of either sliver or roving or yarn are passed through the following different slots (marked on the basis of hank or count) of air capacitors:

SLOT-1	=	0.049–0.294 hank
SLOT-2	=	0.295–3.69 hank
SLOT-3	=	3.7–27.9 Count (Ne)
SLOT-4	=	28–600 Count (Ne)

4.2.5.5   Photo-electric cell type tester

A sample is placed between a source of light and a photoelectric cell, and the variation of light beam generated by yarn unevenness is converted into a variation of electric current which is sufficiently amplified for detection. This method has been tried by many experts [46, 56]. The results obtained by the use of this method are affected by the shape of the cross-section and the way how the fibres are arranged in the yarn, because the quantity measured is proportional to the yarn projection, not to the cross-sectional area as in other methods. Hence, this method is not to be used for slivers and rovings. An apparent variation is often noticed in low twist yarns, because their cross-sections may be flat and appear to have been twisted spirally. A device has been created to prevent this apparent variation due to spiral twisting. It is the Filometer [57], which throws so intense light upon the sample that some parts of light might go through the yarn in order to introduce the effect of the mass.

4.2.5.6   Electric resistance type [48]

This method measures yarn thickness by the electric resistance of the yarn when electric current is transmitted through the sample. The electric conductivity of yarn is generally small, but the weak current can be measured directly or after the sample is immersed in a liquid. In the "after-immersion" system, the degree of the liquid absorption by the yarn should be taken into consideration. The difference of electric resistance of constituent fibres, as well as the homogeneity of chemicals used, will affect the result.

4.2.5.7   Optical method (Zweigle G580)

This instrument measures yarn evenness by a fundamentally different method from the mass measuring system of the Uster instrument. Instead of capacitance measurements it uses an optical method of determining the yarn diameter and its variation. In the instrument an infra-red transmitter and two identical receivers are arranged as shown in Fig. 4.8. The yarn passes at certain speed through one of the beams, blocking a portion of the light to the measuring receiver. The intensity of this beam is compared with that measured by the reference receiver and from the difference in intensities a measure of yarn diameter is obtained. The optical method measures the variations in diameter of a yarn and not in its mass. For a constant level of twist in the yarn the mass of a given length is related to its diameter by the equation:

Mass = $CD2$

where, C = constant,

D = diameter of yarn.

However, in practice the twist level throughout a yarn is not constant [58]. Therefore, the imperfections recorded by this instrument differ in nature from those recorded by instruments that measure mass variation. However, the optical system is claimed to be nearer to the human eye in the way that it sees faults. Because of the way yarn evenness is measured, this method is not affected by moisture content or fibre blend variations in the yarn.

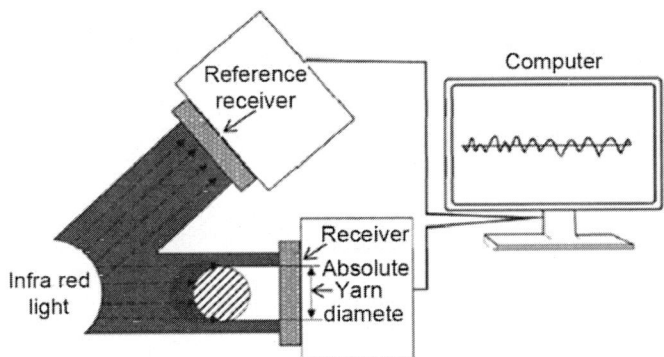

Figure 4.8: Optical method of determining the yarn diameter (Zweigle G580)

4.2.6   Methods of describing evenness

Coefficient of variation

In handling large quantities of data statistically, the coefficient of variation (CV) is commonly used to define variability and is thus well-suited to the problem of expressing yarn evenness. Many authors [62–73] have used it as the basis for developing new ideas or for comparative studies, and it is currently probably the most widely accepted way of quantifying irregularity. It is given by the expression:

$$CV\% = \frac{S}{X}$$

where, S is the standard deviation and X the mean of a number of measured estimates of the mass per unit length of the yarn.

The property is usually expressed as a percentage, calculated as:

$$CV\% = \frac{S}{X}100$$

Irregularity, U%

The parameter U appears to be used almost exclusively by the manufacturers of the well-known Uster test equipment [74]. $U\%$ is the statistical value representing the percentage variation of a yarn. Lower the value the better the yarn, because fewer imperfections exist in the yarn. The average value for all the deviations from the mean is calculated and then expressed as a percentage of the overall mean (percentage mean deviation, PMD).

When the deviations have a normal distribution about the mean the two values are related by the following equations:

$$CV = 1.25PMD$$

$U\%$ is determined on Uster evenness tester on moving average basis. In statistics, a moving average is a calculation to analyse data points by creating a series of averages of different subsets of the full data set. It is also called a moving mean or rolling mean and is a type of finite impulse response filter.

Given a series of numbers and a fixed subset size, the first element of the moving average is obtained by taking the average of the initial fixed subset of the numbers series. Then the subset is modified by shifting forward that is, excluding its first number of the series and including the next number following the original subset in the series. This creates a new subset of numbers and the average is calculated. This process is repeated over the entire data series. The plot line connecting all the fixed average is the "Moving Average". A moving average is commonly used with the time series data to smooth out short term fluctuation and highlight the longer term trends.

Indices of irregularity

The notion of an index to express irregularity appears to have been suggested originally by Spencer-Smith and Todd [75] and have been developed separately by Martindale [76] and Huberty [77]. Briefly, an index of irregularity expresses the ratio between the measured irregularity and the so-called limiting irregularity of an ideal yarn. The manner in which irregularity is assessed can lead to different ways of expressing the index.

In calculating the limiting irregularity, the assumption is made that, in the ideal case, fibre distribution in a yarn is completely random and a practical yarn can never improve upon this situation. Thus, the measured irregularity will be an indication of the extent to which fibre distribution falls short of complete randomness. If all fibres are uniform in cross-sectional size, it can be shown mathematically that the limiting, or minimum, value of irregularity, expressed in terms of CV, is given by:

$$CV_{lim} = \frac{1}{\sqrt{n}}$$

or

$$CV_{lim}\% = \frac{100}{\sqrt{n}}$$

This expression also assumes a Poisson distribution in the values around n, the mean number of fibres in the yarn cross-section, a situation that need not necessarily exist in practice. With these two assumptions, however, an index of irregularity, K, as proposed by Huberty [77] can be derived such that:

$$CV = \frac{K}{\sqrt{n}}$$

Martindale's treatment [76] extends this work on the assumption that, for a practical yarn, variations in linear density of the fibre must be taken into account and modifies the formula for limiting CV according to the equation:

$$CV_{lim}\% = \frac{100\sqrt{1 + 0.00004V_D^2}}{\sqrt{3}}$$

It is an expression that yields typical values for K of 106 for cotton, 112 for wool and 104 for synthetic-staple-fibre yarns [78].

where,

V_D = coefficient of variation of the fibre weight per unit length.

n = the average number of fibres in a cross-section of the strand.

Thus, for a particular fibre and count of yarn, there is limit or basic irregularity upon which our present machinery cannot improve. By calculating the limit irregularity and then measuring the actual irregularity, we have a means of judging the spinning performance.

Let

Vr = the calculated limit irregularity.

V = the actual irregularity.

Then, the *index of irregularity* is:

$$I = V/Vr$$

Addition of irregularity:

In formula given in limit irregularity the square of the coefficient of variation is used; in this form it is known as the "relative variance", often abbreviated to "variance". By using the squares of the coefficients of variation it becomes possible to add and subtract the irregularities produced at various stages in yarn preparation and spinning.

Suppose the coefficient of variation of a sliver is V_1 and it is fed to a machine which adds irregularity to it during processing. Let V be the coefficient of variation of the processed sliver. Using the squares of the coefficients,

$$V_2 = V_1^2 + V_2^2$$

where, V_2 is the coefficient of variation of the added irregularity.

Reduction of irregularity by doubling:

One of the objects of doubling is to reduce the irregularity. If '*n*' strands of material, each having the same coefficient of variation, are doubled, then the coefficient of variation of the combined strands is given by,

CV of doubled strands = CV of individuals/√*n*

4.2.7   Imperfections (thin places, thick places and neps)

Thin places, thick places and neps are part of yarn unevenness. Deviations of the mass from the average yarn body exceeding ±50% are counted as thin and thick places for both ring and rotor spun yarns. Neps (+200% and+280% for ring and rotor spun yarns, respectively) are short thick places resulting from fibre entanglements of frequently immature fibres or seed coat fragments. Imperfections in the cross-section of yarn will heavily increase with higher yarn count, i.e. with fewer fibres in the cross- section.

The higher the short fibre content, the higher the number of imperfections. Frequent imperfections can be very disturbing in a fabric. Fibre entanglement often results from immature fibres which cannot absorb dyestuff and, therefore, remain white.

Thin place

These are number of places that have mass reduction of 50% or more with respect to the mean value. Note that (−50%) is the standard sensitivity level used in the test. If a different sensitivity level (−30%, −40%, −60%) is used, the result would have been different. These thin places have a length of approx. 4.0 mm.

Thick place

These are number of places that have mass increase of 50% or more with respect to the mean value. Note that (+50%) is the standard sensitivity level used in the test. If a different sensitivity level (+35%, +70% and +100%) is used, the result would have been different. These thick places have a length of approx. 4.0 mm.

Neps

These are number of places that have mass increases of +200% or more for ring spun yarns and +280% or more for rotor spun yarns with respect to the mean value and a reference length of 1 mm. Note that +200% and +280% are the sensitivity levels normally used in the test. These short thick places in a yarn are often the results of vegetable matter or entangled fibres.

Table 4.3 describes different types of imperfections at different levels of variation.

Table 4.3: Imperfection levels

Thin	Types
(−60% level): It means cross-section of thin places is only 40% of the yarn mean value or less	Serious thin places, easy to detect on yarn's black board from a few meter distance
−50%: Its means cross-section of thin places is only 50% of the yarn mean value or less	Quite serious, visible from one meter distance
−40%: It means cross-section of thin places is only 60% of the yarn mean value or less	Smaller thin places visible on a black board from a close distance
−30%: It means cross-section of thin places is only 70% of the yarn mean value or less	Very small thin places, black board shows the fault

Thick	Types
+70%: It means cross-section of thick places is +170% of the yarns mean value or more	Quite serious thick places visible on black board from a distance of few meters
+50%: It means cross-section of thick places is +150% of the yarn mean value or more	Smaller thick places visible on black board at a close distance
+35%: It means cross-section of thick places is +135% of the yarn mean value or more	These places hardly visible on black board

Neps	Types
+400%: Cross-section of neps is 500% of the yarn mean value	Very large nep
+200%: Cross-section of neps is 300% of the yarns mean value	Smaller neps detectable on black board from a close distance
+140%: Cross-section of neps is 240% of the yarns mean value	Very small neps detectable on black board from a very close distance

Difference between neps and thick places:

Neps are considered as thick places whose length is shorter than 4.0 mm. Thick places are longer than 4.0 mm.

We check the yarn's imperfection at the following levels which are spun on cotton system:

Thick   –   +50%,
Thin   –   −50%,
Neps   –   +200%.

Extra sensitive imperfection is checked at the following levels.

Thick   –   +35%,
Thin   –   −40%,
Neps   –   +140%.

4.2.8   Graphical representation of mass variations

Graphical representations are aimed at providing easy analysis possibilities as well as providing more complete information than the numerical estimates. The following graphical representations are common with the latest generation evenness testers:

- Spectrogram,
- Variance length curve,
- Normal diagram,
- Cut length diagram.

4.2.8.1   Spectrogram

The numerical values such as $U\%$ or CV% are not influenced by the periodic variations. A spectrogram is used to measure the periodic or nearly periodic mass variations in a sliver, roving and yarn by analysing the frequencies at which faults occur electronically.

The spectrogram (or spectrograph) is a graphical representation specifically designed for identification and analysis of the periodic faults. It is a representation of the mass variations in the frequency domain. In other words, a spectrogram shows how many times a mass variation repeats itself in a tested length of yarn. Figure 4.9 shows an example of the spectrogram for a yarn.

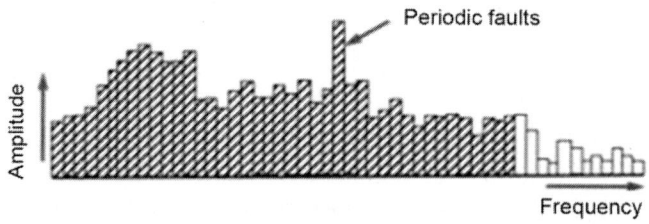

Figure 4.9: Frequency spectrums showing periodic fault

4.2.8.1.1 Arrangement of spectrogram

In a spectrogram, the X-axis represents the wavelengths and Y-axis represents the amplitude of the faults as shown in Fig. 4.9. Often a logarithmic scale is given for the X-axis to cover the maximum range of wavelengths.

The spectrogram consists of shaded and non-shaded areas. If a periodic fault passes through the measuring head for a minimum of 25 times, then it is considered as significant and it is shown in the shaded area. Wavelength ranges which are not statistically significant are not shaded. In this range the faults are displayed but not hatched. This happens when a fault repeats for about 6 to 25 times within the tests length of the material. As far as those faults in the unshaded area is concerned, it is recommended to first confirm the seriousness of the fault before proceeding with the corrective action. This can be done by testing a longer length of yarn. Faults which occur less than 6 times will not appear in the spectrogram.

A spectrogram starts at 1.1 cm if the testing speed is 25–200 m/min. It starts at 2.0 cm if the testing speed is 400 m/min and it starts at 4 cm if the speed is 800 m/min. For spun material the maximum wavelength range is 1.28 km.

Wavelength range covered by an evenness tester depends on the test speed and the evaluation time. With staple fibre yarns it covers a wavelength range from 2 cm to 1280 m and with filament yarns a range of 2 cm to 2560 m.

Spectogram at different processing stages are shown in Fig. 4.10.

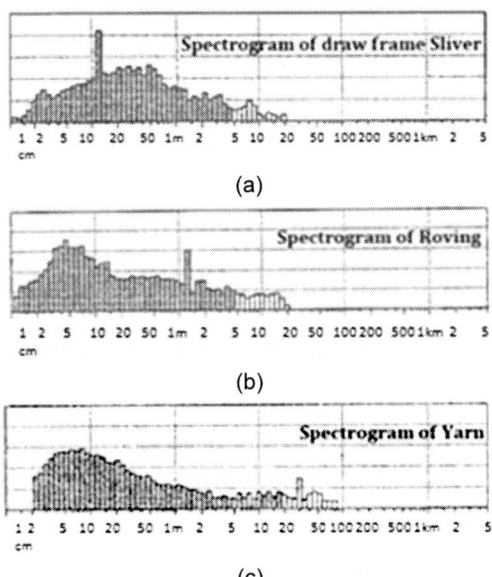

(a)

(b)

(c)

Figure 4.10: Spectogram at different processing stages. (a) Draw frame sliver, (b) Roving and (c) Yarn

4.2.8.1.2 Theoretical spectrogram

If the CV of a yarn is zero then the spectrogram consists of a straight line. If the yarn has a completely random distribution of staple fibres, the staple length L has an effect on the spectrogram.

Case 1: The spectrogram of a fault free yarn consisting of all fibres with equal length (e.g. staple fibre yarn) will be as shown in Fig. 4.11(a). It will be started at zero point corresponding to the staple length and a maximum value at 2.7 times the staple length.

Case 2: The spectrogram of a fault free yarn consisting of natural or variable length fibres will be as shown in Fig. 4.11(b). In this spectrogram, the maximum amplitude lies at a wavelength of 2.82 ×mean fibre length.

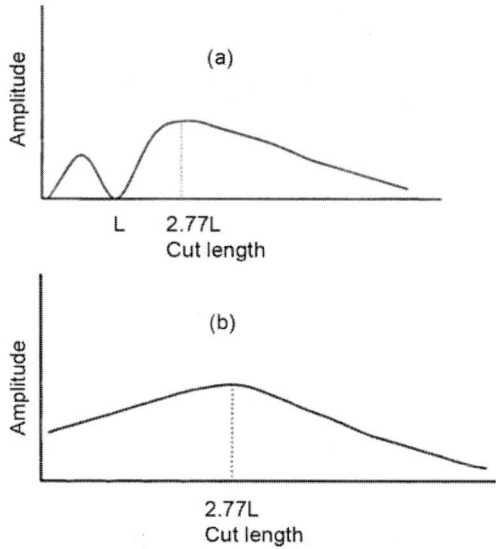

Figure 4.11: The spectrogram of a fault free yarn.
(a) Fibres with equal length and (b) Fibres with variable length

4.2.8.1.3 Spectrogram with common periodic faults

Mainly two types of periodic faults are shown in the spectrogram (Fig. 4.12). They are as follows:

    1. Hill type faults.

    2. Chimney type faults.

Hill type faults

A hill type fault, where several adjacent peaks are noticed, is normally due to drafting waves caused by factors such as improper draft zone settings, improper top roller pressure, too many short fibres in the material, etc. The wave length at which maximum amplitude of the hill occurs is called as the average wave length. Drafting faults are created and influenced by non-optimal settings of one or several of the following factors:

- Gauge distance between the drafting rollers (Nip).
- Roller pressure.
- State of the roller's surfaces.

When searching to eliminate drafting faults, one would look for the main cause in one of those factors first. In many cases though, a compromise has to be found, since certain materials are more critical. Example: Combed cotton draw frame slivers, where the fibres are highly parallel and thus slippery and difficult to draft optimally at a reasonable speed, a drafting fault hill is to be found at a wavelength of about 2.8 × average fibre length. If the drafting fault hill does not lie around 2.8 × average fibre length, one has to divide the wavelength λ of the hill crest by 2.8 × average fibre length in order to get the approximate draft factor back to the origin of the fault.

Formula:

$$\text{Draft ration} = \frac{\lambda^{crest}}{2.8 \times \text{average fibre length}}$$

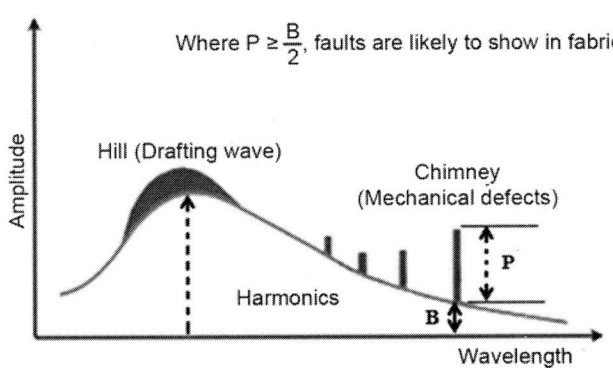

Figure 4.12: Spectrogram with common periodic faults

Chimney type faults

A chimney type fault, consisting of one or two 'peaks' or 'chimneys', is normally due to a mechanical fault such as eccentric roller/gear, improper meshing, missing gear teeth, missing teeth in the timing belts, damaged bearings, etc.

If the height of the peak (P) above the basic spectrogram at any wavelength equals or oversteps by 50% of the height of the basic spectrum at that wavelength i.e. $P \geq B/2$, then it can be considered to be sufficiently serious fault. A chimney would be formed only if the wave length of periodic fault is detected minimum 25 times for the tested length. No chimney will be formed for length repeating less than six times in tested length.

Peaks or chimney occurring in non-shaded portions are having no statistical significance. In case peaks do occur, they can be magnified by running the same material at a different range of test, longer time of test.

Harmonics wave

Periodic variation is sinusoidal i.e. gradually decreasing and increasing in size. But there are certain periodic variations which are not sinusoidal. In many cases, a single periodic material fault produces multiple chimneys. Multiple chimneys are the result of a periodic yarn mass variation which is not evenly shaped, i.e. not sinusoidal. A multiple periodic fault consists of a base wavelength and of so-called harmonic wavelengths. The harmonics are usually to be found at factor 1/2, 1/3, 1/4, etc. of the base wavelength. The longest wave length which appears in the spectrogram is the main culprit which is called fundamental wave. Corrective action on fundamental wave helps to remove other peaks also.

4.2.8.1.4 Procedure for drawing the reference line

Mark "dot" at the corner of the rectangular bar and join the dotted marking by drawing curved lines through them. Some or few dotted portion may get omitted for maintaining the shape of the reference line as shown in Fig. 4.13.

For better representation of the spectrogram, the reference line could be drawn by "blue" felt pen, the chimney by "red" and drafting wave by "green".

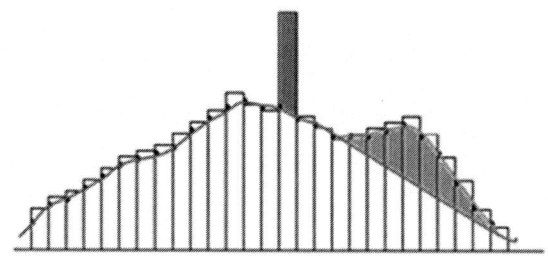

Figure 4.13: Procedure for drawing the reference line

4.2.8.1.5 Different information available from spectrogram (UT-3)

1. Type of material tested (sliver/roving/yarn):

This can be read out from the chart as follows:

a. Chart starts from 1.1 cm and terminate at 1.3 m (shaded portion) with a material speed of 25 m/min – sliver.

b. Chart starts from 1.1 cm and terminate at 2.3 m (shaded portion) with a material speed of 50 m/min – roving.

c. Likewise, starts from 2.3 cm and terminate at 18 m (shaded portion) with a material speed of 400/min – yarn.

2. Type of fibre (cotton/wool or synthetic):

By looking into the shape of reference line of the spectrogram, as shown in Fig. 4.14, it is possible to identify the type of fibre present in the tested strand.

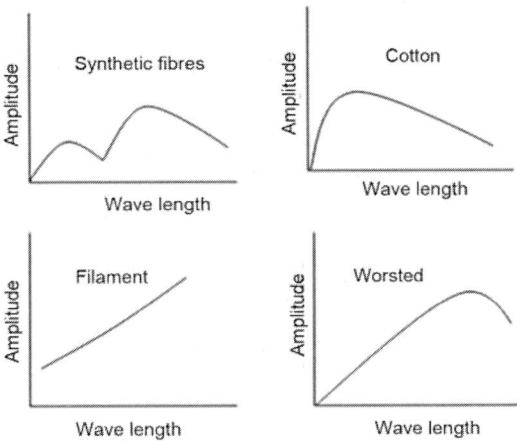

Figure 4.14: Shape of reference line of the spectrogram for different types of fibre

3. Nature of mass variation:

Nature of mass variation can be detected by observing the steepness of the reference line, as shown in Fig. 4.15.

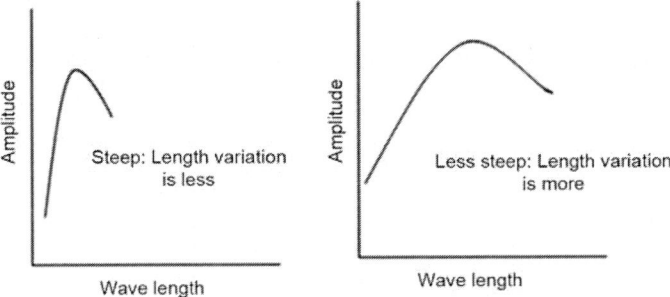

Figure 4.15: Steepness of the reference line

4. Fibre length distribution:

Fibre length distribution can be ascertained by observing the inclination of the reference line with the X-axis, as shown in Fig. 4.16. Higher the angle, lesser is the mean staple length (MSL).

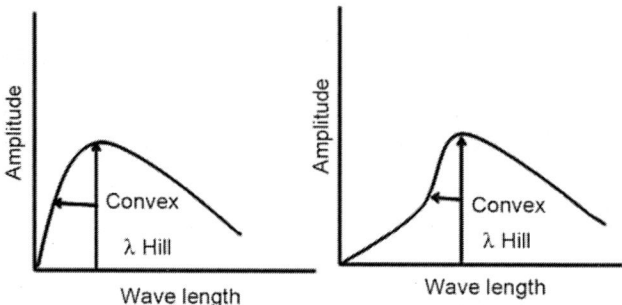

Figure 4.16: Inclination of the reference line

5. Whether the yarn tested is carded or combed:

This can be predicted by looking into the shape of the reference diagram, as shown in Fig. 4.17. Carded yarn is depicted by a convex shape. Here mean staple length (MSL) is less. Combed yarn is depicted by concave shape and the mean staple length (MSL) is more.

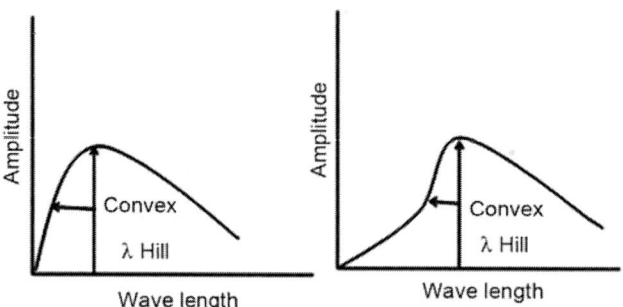

Figure 4.17: Shape of the reference line

6. Mean staple length (MSL) of the mixing:

Mean staple length (MSL) of the mixing can be calculated from the spectrograph of sliver/roving/yarn. MSL is related with the wavelength of highest amplitude of hill type faults (λHill) as follows:

  (i) For Yarn, λHill = 2.75 × MSL.

  (ii) For Roving, λHill = 3.5 × MSL.

  (iii) For Drawing, λHill = 4.0 × MSL.

7. To identify the nature of fault – mechanical or drafting waves:

Periodic variation caused by mechanical faults is characterised by one particular wavelength. The thick and thin places occurring in one wave length go on repeating over the length of the strand. A periodic fault is shown in the spectrogram as a distinct peak with one or two channels or chimneys.

Periodic variation is primarily due to any one of the following reasons:

Eccentric rollers, damaged aprons, defective gears, improper meshing, worn out bearings and damaged surface of the rollers.

Drafting waves are represented in spectrogram in the form of "hill" spanning over many channels. The wave length at which the maximum amplitude of the hill occurs is called as the average wave length.

The basic cause for the formation of drafting waves is "floating fibres". Floating fibres are fibres which are not gripped by any roller pairs in the drafting zone. Also, improper roller setting, improper roller pressure, high drafts, etc. increase the amplitude of drafting wave.

4.2.8.1.6 Spectrogram analysis for finding source of fault

How to analyse the spectrogram for periodic faults:

Following few points must be kept in mind while taking spectrogram:

1. While taking spectrogram of any machine like drawing/simplex/ring frame, etc., the drafting zone, top rollers, bottom rollers, feed table, etc. are to be cleaned from fluff and flies. Presence of fluff/flies on the above areas may disturb the alignment of fibres assembly and will cause false peaks.

2. After taking above precaution, if peak/wave appears, then repeat for 2nd time for confirmation.

3. Suppose any faulty gear is pin pointed through spectrogram, it is better to look into the train of gears in that region.

4. Presence of any loose fluffs/flies, hardened grease accumulated inside the tooth/teeth of any gear must be cleaned.

Calculation of wave length for defect in different relevant parts:

Make a readymade table as given in Tables 4.4 and4.5 from each type of machine to find out the wavelength (λ) with its corresponding part responsible for giving peak. Calculation should start from delivery/front roller side and then going to the back roll or feed roll.

Ring frame:

Gearing diagram of a ring frame taken for calculation is as shown in Fig. 4.18.

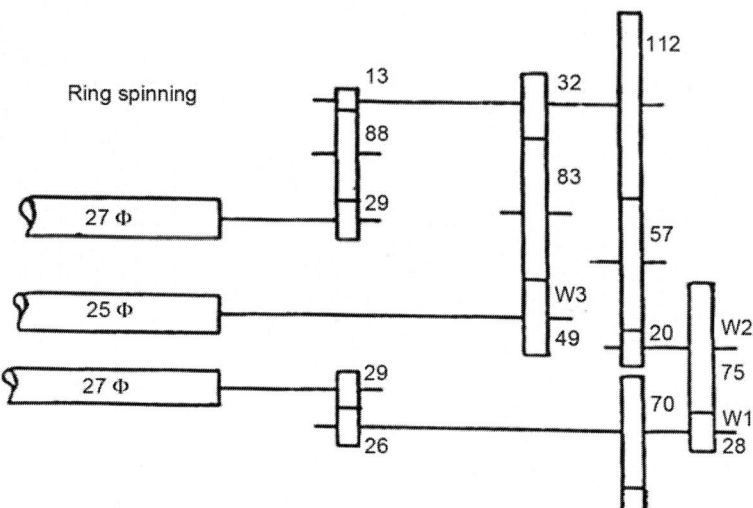

Figure 4.18: Gearing diagram of a ring frame drafting system

Table 4.4: Wave length of periodic fault caused from different source in ring frame

Sl. no.	Machine part	Wavelength (λ) (cm)	Source of periodic fault
1.	$(27 \times 3.14)/10$	8.48	27 mm front roll
2.	$8.48 \text{ cm} \times \dfrac{26}{29}$	7.6	26 T Gear
3.	$7.6 \text{ cm} \times \dfrac{75}{28}$	20.35	75 T Gear
4.	$20.35 \text{ cm} \times \dfrac{57}{20}$	58	57 T Gear
5.	$58 \text{ cm} \times \dfrac{112}{57}$	114	112 T Gear
6.	$114 \text{ cm} \times \dfrac{83}{32}$	296	83 T
7.	$296 \text{ cm} \times \dfrac{49}{83}$	175	25 mm middle roller
8.	$114 \text{ cm} \times \dfrac{88}{13}$	772	88 T Gear
9.	$772 \text{ cm} \times \dfrac{29}{88}$	254	27 mm back roller

Simplex (LF 1400A)

Gearing diagram of Simplex (LF 1400) taken for calculation is shown in Fig. 4.19.

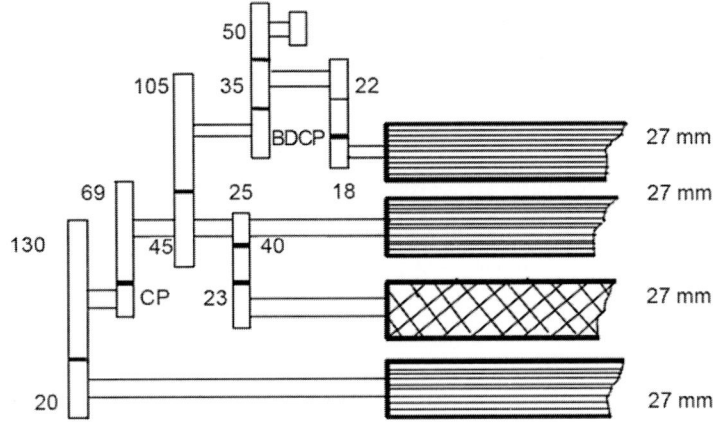

Figure 4.19: Gearing diagram of a Simplex drafting system

Table 4.5: Wave length of periodic fault caused from different source in simplex

Sl. no.	Machine part	Wavelength (cm)	Source of periodic fault
1.	2.7 cm × 3.14 cm	8.48	27 mm Front roller
2.	$\dfrac{8.48 \text{ cm} \times 130}{20}$	55	130 T gear
3.	$\dfrac{55 \text{ cm} \times 69T}{CP = (50T)}$	76	69TBRW/3rd roll 27 mm
4.	76 cm × 23/25	70	27 mm Bottom fluted 2nd roll
5.	76 cm × 105/45 cm	177	105 T/BDW
6.	177 cm × 35 /BDW (=57 T)	109	35 T Gear
7.	$109 \text{ cm} \times \dfrac{18}{22} \text{ cm}$	89	Back bottom fluted roll 27 mm

Shortening of the wavelength in spectrogram

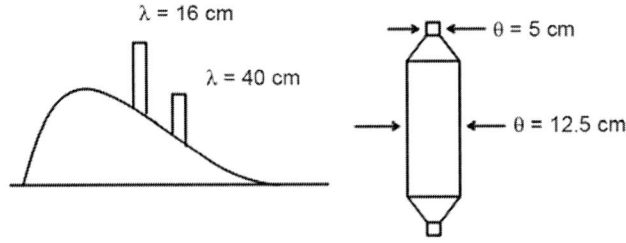

Figure 4.20: Spectrogram taken from rovings

Faults appeared in spectrogram taken from rovings.

$\lambda = 16$ cm Ø = 5 cm

$\lambda = 40$ cm Ø = 12.5 cm

i.e. wavelength of the periodic variation correspond to the circumference of the package produced.

1. The sources of the above faults are due to:
2. Eccentric spindle in the simplex.
3. Eccentric running of the bobbin.
4. Eccentricities between spindles and flyers of the speed frame.

4.2.8.2. Variance length curve

The variance length curve is widely acknowledged to be an important and useful method of expressing quantitatively the evenness of a fibre assembly. It is capable of displaying at a glance the presence of non-periodic faults and of illustrating the relative proportions of medium and long-term irregularities in the test material. For this reason, several methods of constructing the curve have been devised, normally in conjunction with electronic evenness testers [79–86].

Tippett [87] and Martindale [88] have all considered different aspects of the relation between variance and the length of yarn measured. Tippett gives a theoretical discussion of the effect of length measured on the variance between lengths of cotton slubbings, etc., and Martindale quotes results for the same effect in wool. Study made by Townsend and Cox [89] is based on the relation between a length L of yarn and the mean standardised variance, $V(L)$, within random samples length L. The "mean standardised variance" is simply the square of the coefficient of variation, and is a more useful index for comparative purposes than the mean variance $V(L)$.

If $B(L)$ is defined as the standardised variance between the means' of lengths L of yarn, then from the simple theory of analysis of variance:

$$V(L) + B(L) = V(\infty)$$

where, (∞) is the over-all variance.

The general shapes of the $V(L)$ and $B(L)$ curves are given in Fig. 4.21 and we shall consider some of the parameters associated with the $V(L)$ curve that can be directly connected with the properties of the yarn.

It is quite possible that one yarn will have a relatively large amount of short term irregularity and not very much long term irregularity while a second will have a relative amount of short and long term irregularity reversed. The overall irregularity in terms of CV may be the same for the yarn, but the effect in the fabric may be quite different [90].

The general shapes of the $V(L)$ and $B(L)$ curves are given in Fig. 4.21 and we shall consider some of the parameters associated with the $V(L)$ curve that can be directly connected with the properties of the yarn.

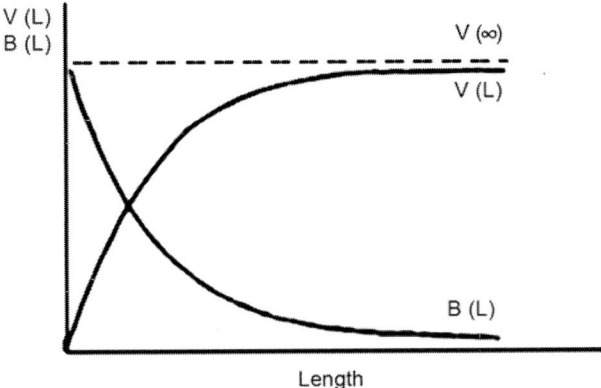

Figure 4.21: General shapes of the V (L) and B (L) curves

4.2.8.2.1. Asymptotic value $V(\infty)$

When L is large, greater than about 1 m, $V(L)$ increases only slightly with large increases in L, and we may assume that it approaches an asymptotic value $V(\infty)$. Most determinations of the variance in a yam refer to lengths greater than 2 or 3 m and so give values near to $V(\infty)$.

(a) Rapidity of approach to $V(\infty)$

Consider two yarns A and B with the same $V(\infty)$ and suppose that V is almost equal to $VA(L)$ when $L = 1$ m, whereas at this value of L, $VB(L)$ is appreciably less than the asymptotic value (Fig. 4.22). It is clear that these two yarns will have different appearances and should be distinguished in any comparative investigation. We may say that yarn B has a "long term" variation not present in yarn A. This difference is detected as a difference between the rates at which the two curves approach $V(\infty)$.

(b) Gradient at the origin

Having considered the behaviour of $V(L)$ for large values of L, we now consider the interpretation of $V(L)$ for small values, say 0.5 cm. Over this relatively small range $V(L)$ may be taken as linear, and its behaviour described by the gradient of the $V(L)$ curve at the origin.

Consider two yarns C and D with the same rapidity of approach to a common $V(\infty)$, and suppose that $V(L)$ is effectively constant for L greater than 1 m. Yam C has a high gradient at the origin, where yam D has a low gradient (Fig. 4.23). We can see that the mean standardised variance within 1 cm. lengths is much greater for yarn C than for yarn D. It is not clear which of these will be the better yarn, but their properties are not identical.

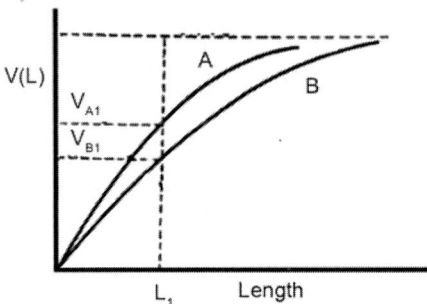

Figure 4.22: Rapidity of approach to V (∞)

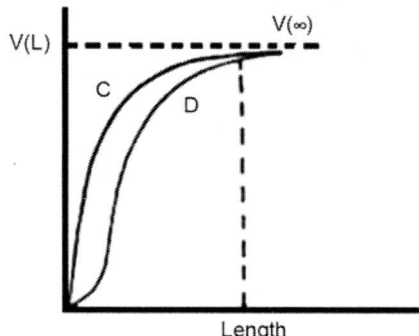

Figure 4.23: Gradients at the origin

(c) Log–log plot of variance length curve

In capacitance type evenness tester like Zellweger Uster, the variance length curve is produced by calculating the CV for different cut lengths and plotting it against the cut length on log–log paper. A perfect yarn would produce a straight line plot. The curve is a useful tool for examining medium and long-term non-periodic variations in a yarn. The better is the evenness of the yarn the lower is the curve and the steeper is the angle it makes to the cut length axis. This is shown in Fig. 4.24, where the variance length curve for an actual cotton yarn is compared with a curve for an ideal yarn. The measured curve deviates from the theoretical curve in the region where there is long-term variation in the yarn. The variance length curve of a poor fibre assembly lies above the curve of a good fibre assembly as shown in Fig. 4.25 where the poor yarn diverges from the good yarn at the longer cut lengths.

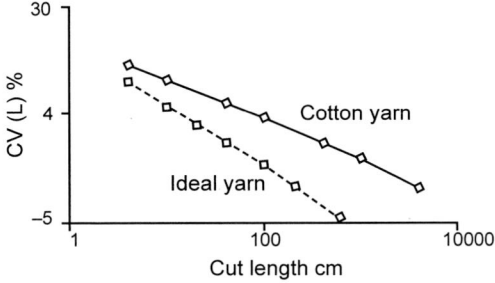

Figure 4.24: Variance length curves for cotton and ideal yarns

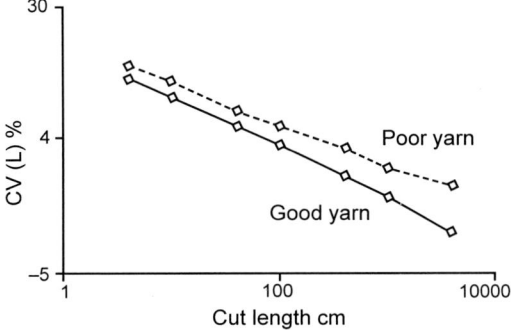

Figure 4.25: Variance length curves for poor and good yarns

(d) Variance–length (V–L) curve and its interpretation

It is a meaningful exercise of QA department to analyses the V–L curve apart from $U\%$ and imperfection. How can it be done is discussed in details here.

Uster uneveness tester determines over all irregularity i.e. it determines CV of yarn according to law of statistics. With these values, we can determine the small differences with respect to mass variation. Attempts were made by various research workers for a detailed analysis of mass variation by calculating CV values at different cut length. The outcome of this study was the variance length curve abbreviated as (V–L) curve.

Let us consider a long length of yarn 'L' is cut into a number of pieces as $l_1, l_2, l_3, \ldots, l_n$.

Suppose CV of $l_1 = \mathrm{CV}_1$, CV of $l_2 = \mathrm{CV}_2 \ldots$ CV of $l_n = \mathrm{CV}_n$ then, $\left(\dfrac{\Sigma \mathrm{CV}_n}{n} \right)^2$

is called variance length (V_L) within lengths and $\{\mathrm{CV}\,(L)\}^2$ is called variance length (BL) between lengths.

Unevenness tester actually gives us CV of cut length. What we actually get from evenness tester is the variance length (V–L) curve obtained by plotting CV of different cut lengths on a double logarithmic paper with CV (L) on Y-axis and cut length on X-axis, as shown in Fig. 4.26.

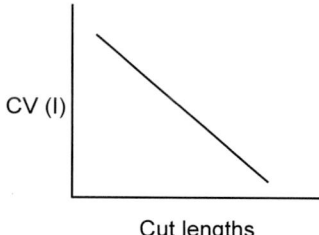

CV (l)

Cut lengths

Figure 4.26: Variance length curve on log–log plot

This curve is a straight line instead of a curve line as it is drawn on a double logarithmic paper, as obtained from Uster evenness tester.

The yarn which is free from any variation will appear as a straight line and any bent or deviation from the straight path depicts the presence of variation which may be medium or long term according to the position of the bent portion. If the bent is at the start of the curve, it is responsible for giving streaks in the fabric and if the deviation occur at the end of the line then it is long term variation and responsible for giving patta in the fabric, as shown in Fig. 4.27.

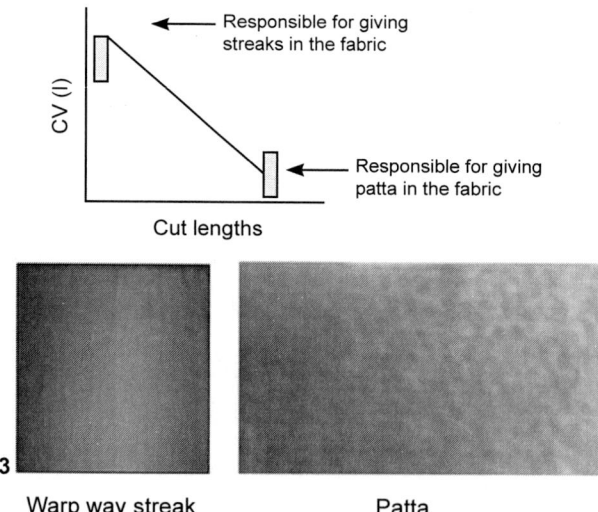

CV (l)

Responsible for giving streaks in the fabric

Responsible for giving patta in the fabric

Cut lengths

3

Warp way streak   Patta

Figure 4.27: Significance of bent of the variance length curve

The angle of inclination of the line with X-axis gives the information on the nature of mixing. Lesser inclination angle represents a richer mixing and a higher inclination angle represents a poorer mixing. If two types of mixing are taken e.g. (i) richer, (ii) poorer and (V-L) curve for both the yarns are taken on a same graph sheet, the (*V–L*) curve of poorer mixing shall lie above the richer mixing in (*V–L*) curve as explained in Fig. 4.28.

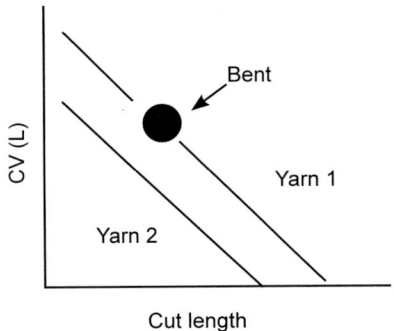

Figure 4.28: Angle of inclination of the (V–L) curve

Yarn (1) – Poorer mixing with a process fault at bent portion.

Yarn (2) – Richer mixing with an angle of inclination lesser than yarn (1).

Thus, (*V–L*) curve gives the following information:

(1) Process fault at different processing stages of spinning which is non-periodic.

(2) Nature or type of mixing (for cotton).

(3) Type of impact of yarn's fault in fabric i.e. streak effects or patta effect.

(4) Type of variation i.e. whether medium/long.

How to find out which spinning stage is introducing faults in the yarn

Malatinszky and Grosberg [91] described a relation between cut lengths and mean fibre length as follow:

$$l^* = l\,(1 + CV_1^2) \qquad\qquad ...(1)$$

where,

l^* = initial cut length for calculating the cut length referring to a process stage,

l = mean fibre length (mm).

Equation (1) can be written as:

$$l^* = lK \text{ where } K = 1 + CV_l^2$$

The value of K for cotton = 1.18, wool = 1.27, synthetic fibre = 1.0.

In case of a blended yarn like polyester/cotton (80/20), the value of K can be calculated as follows:

Take, mean length of cotton = 20.5 mm = l_e (say),

Take mean length of polyester = 32 mm = l_p (say),

Mean length of the blended yarn = 20.5 × 0.20 + 32 mm × 0.80 = 29.7 mm.

Now, for blended yarn, $\sigma^2 = \sigma_e^2 + \sigma_p^2$ ...(2)

where,

σ_e = standard deviation of cotton fibre length,

σ_p = standard deviation of polyester fibre length,

σ = standard deviation of blended fibre length.

Equation (2) can be written as:

$$\sigma_2 = cv_c^2 \cdot l_c^2 + cv_p^2 \cdot l_p^2 \qquad ...(3)$$

$(s = cv \times l)$

The value of cv_c^2 and cv_p^2 can be found out from the relation $k = 1 + cv^2$.

For cotton, $cv_c^2 = 1.18 - 1 = 0.18$ [$K = 1.18$ for cotton],

For polyester, $cv_p^2 = 1.0 - 1.0 = 0$ [$K = 1.0$ for synthetic fibre].

$l_c = 20.5$ mm, $l_p = 32.0$ mm

From Equation (3), $\sigma_2 = (0.18) \times (20.5)^2 + 0 \times (32.0)^2 = 75.64$ mm.

Now, from the relation, $K = 1 + cv^2$, we can write $K = 1 + (\sigma/l)_2$.

Here $\sigma^2 = 75.64$ mm.

$l^2 = (29.7)^2$ mm.

So, $K = 1 + 75.64/(29.7)^2 = 1.0858$ is the final value of K for 80/20 polyester/cotton blended yarn. This value of K is the primary requirement for locating the faulty process stage.

Suppose mechanical draft at different process stages are:

Ring frame stage = D_1 (say),

Simplex frame stage = D_2 (say),

Drawing frame stage = Finisher = D_3 (say)

= breaker = D_4 (say)

With the help of this relation as $l^* = lK$, all process stages can be determined as:

(i) Cut length range for simplex = $l^* \times D_1$.

(ii) Cut length range for drawing(F) = $l^* \times D_1 \times D_2$.

(iii)   Cut length range for drawing$(B) = l^* \times D_1 \times D_2 \times D_3$.

(iv)   Cut length range for carding $= l^* \times D_1 \times D_2 \times D_3 \times D_4$.

Now from $(V\text{--}L)$ curve, locate the deviation point from the straight line and find out the cut length corresponding to the deviation or broken point. Now verify this cut length with any one of the above cut length range. Suppose, it come on drawing (F). Then, drawing (F) is introducing the fault in the yarn.

Example: Count $= 30$ scarded, mixing $= 100\%$ cotton,

Total ring frame bobbin taken 10.0.

Now, we know the relationship $l^* = Kl$, where, $K = 1.18$. Take spectrograph of the yarn and find out λ_{Hill} or λ_{max}. Suppose it came at 60 mm.

$$\text{Mean length} = \lambda_{\text{Hill}} \text{ or } \lambda_{\text{max}}/2.75 = \frac{60 \text{ mm (say)}}{2.75} = 21.8 \text{ mm.}$$

So, $l^* = lk = 1.18 \times 21.8 \text{ mm} = 2.57 \text{ cm.}$

Now, draft distribution at different process stages are:

$$\text{Ring frame draft } (D_1) \ = \ 30.0,$$
$$\text{Simplex frame draft } (D_2) \ = \ 7.0,$$
$$\text{Drawing } (F)(D_3) \ = \ 7.9,$$
$$\text{Drawing } (B)(D_4) \ = \ 8.16.$$

So,

Cut length range for simplex $= l^* \times D_1 = 2.57 \text{ cm} \times 30 = 77.10 \text{ cm.}$

Cut length range for drawing $(F) = l^* \times D_1 \times D_2 = 77.1 \times 7 = 5.40 \text{ m.}$

Cut length range for $(B) = l^* \times D_1 \times D_2 \times D_3 = 5.40 \times 7.9 = 42.66 \text{ m.}$

Cut length range for Carding $= l^* \times D_1 \times D_2 \times D_3 \times D_4 = 42.66 \times 8.16 = 348.11 \text{ m.}$

Now, from $(V\text{--}L)$ curves of 10 ring frame bobbins, find out the broken deviation points from straight line of 10 bobbins. Suppose, out of 10 bobbins, at least 4 bobbins are showing the deviation point at the cut length of drawing (B). Then, drawing (B) is the probable source for introducing this long term variation in the yarn.

4.2.8.3   Normal diagram

The normal diagram represents a graphical plot of the basic mass variations of the textile material over its length. The reference length for the mass values is the basic measured length of 1 cm. The diagram traces out the mass variation of fibres assembly with reference to time as shown in Fig. 4.29.

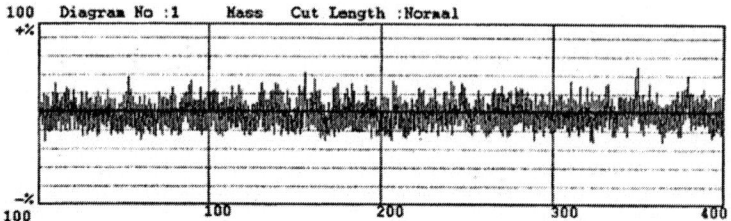

Figure 4.29: Mass variation of fibres assembly with reference to time

The diagram provides additional information on the mass variations which cannot be obtained with other forms of representation – either numerical or graphical. The following are some of the information which can be usefully applied for process control:

- both periodic and non-periodic variation in the material.
- extreme thick places/thin places.
- slow changes in the mean value.
- step changes in the mean value.

Difference between spectrogram and diagram is described in Table 4.6.

Table 4.6: Difference between spectrogram and diagram

Spectrogram	Diagram
Can be used for locating and correcting mass variation occurring at specific wave length	Can mainly be used for analysing trends of short and medium term variation
It is the representation of the mass variation in the frequency domain	It is the representation of the mass variation in time domain

Cut length diagram

The cut length diagram is a graphical representation similar to a normal diagram but represents mass variations over specified cut lengths like 1, 3, 10,

100 m, etc., as shown in Fig. 4.30. The variations over lengths lesser than the cut length of the diagram are suppressed.

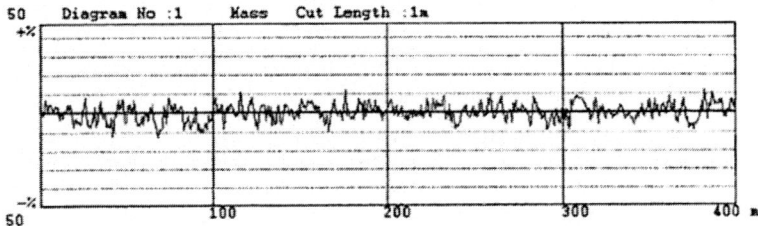

Figure 4.30: Cut length diagram

With the help of diagrams of different cut lengths, the following information can be obtained:

- Checking of the functioning of auto leveller fitted draw frames.
- Checking of count variation (with cut length 100 m).
- Setting of the sensitivity of the count channel in the latest generation Yarn Clearers.

4.2.9   Yarn faults

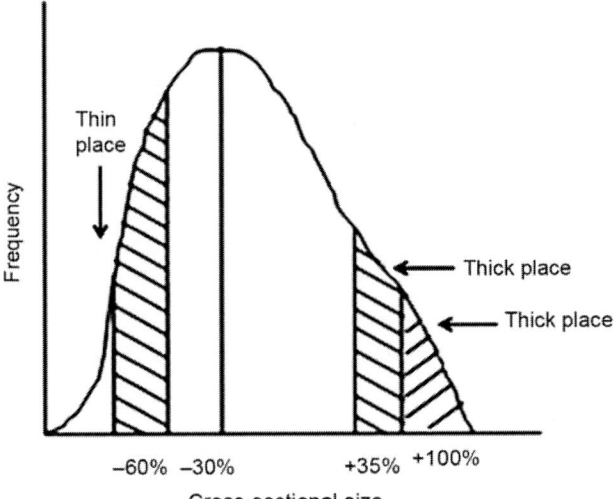

Figure 4.31: Frequency distribution of yarn's cross sectional sizes

Apart from frequently occurring faults (imperfections), there are also seldom occurring faults which include slubs, fly piecing, etc. having specific dimension in terms of diameter and length. Classimat faults are termed as seldom occurring faults as they occur in more range length of yarn whereas imperfection occur is shorter range length of yarn which is termed as frequent faults. These faults are different from imperfection by their sizes in terms of faults length and cross sectional sizes. Thick faults are more than 100% of average weight per unit length of yarn where as thick places in imperfection lie between +35% and+100% of the average weight per unit length of yarn. Frequency distribution of yarn's cross sectional sizes are shown in Fig. 4.31.

These faults are more visible over a long length of yarn and do not impair the appearance of the fabric to that extent as frequent faults do, but downgrade the value of the fabric as per presence of faults. Classimat faults are expressed per 100 km of yarns.

Locher and Ernst [92] mentioned that faulty places which are most difficult to be kept in control during spinning are the relatively seldom occurring yarn faults. These include on one hand thickening in the yarn of different length but with a size which is multiple of the cross-section of the normal yarn and on the other hand long thin place whose cross section are reduced to less than half of normal yarn cross sections. Measurement of seldom occurring faults is not a very common phenomenon in the textile industry. This is so because the length of yarn that needs to be tested to count these types of faults is very large up to a minimum of hundred kilometres. Despite testing of very huge quantity of yarn the frequency of occurrence of some of the faults is still very less to draw any meaningful conclusion for one to take specific steps of corrective action. However, studies carried out on this subject show that seldom occurring faults like slubs, spun in lint, loose lint, piecing, long thick place, long thin place, etc., have a significant contribution to end breaks, during spinning or more so in the subsequent processes. These faults if pass unbroken, are also many times disturbing or objectionable in the fabric leading to rejection of the fabric.

4.2.9.1   Measurement of yarn faults
There are basically two principles by which yarn faults can be measured:
- capacitive method,
- optical method.

4.2.9.1.1 Capacitive methods
This method of detecting yarn fault is based on the principle of measuring thickness of yarn by measuring capacitance as already described in case of capacitance type yarn evenness testers.

They measure the fault and give the output in terms of cross sectional size and length. Faults are classified into different size and length groups and shown in the form of a matrix.

This will provides a detailed break-up of yarn faults based on the length and thickness of the faults, as shown in Fig. 4.32. Uster classimat II gives 23 types of faults, each fault is identified by an alphabet and a number. The alphabet 'A' to 'G' indicates the thick yarn faults while the letters 'H' and 'I' represent thin faults. The number 1 represents a fault of size +100% to +150% over the nominal cross-section in the case of thick place and −30% to −45% in the case of thin places. Number 2 means +150% to +250% for thick places and −45% to −75% for the thin places and so on, as shown in Fig. 4.32.

The 'E–G' categories of faults are not followed by any numerical since they are fewer in number.

Uster classimat III classifies faults into 33 classes. In addition to classimat II faults there are very short thick faults (A0, B0, C0, D0) and short thin faults (TB1, TC1, TD1, TB2, TC2, TD2) as shown in Fig. 4.32.

Clsssimat II is widely accepted by the industry because the 23 faults detected in classimat II are good enough to cover the range of yarns normally produced by the industry. The extended 10 classes in classimat III are not used for normal purpose and mainly considered in some specific cases and for ply yarns.

Classification diagram

Uster classimat II   **Uster classimat III**

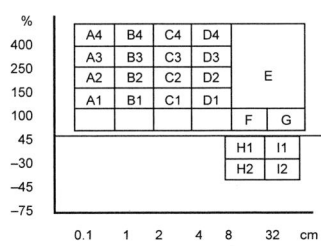

Length classes:
A. Shorter than 1 cm
B. 1to 2 cm
C. 2 to 4 cm
D. 4 to 8 cm
E. Larger ghan 8 cm
F & H. 8 to 32 cm
G & I. More than 32 cm

Cross-section classes:
1. +100 to + 150%
2. +150 to +250%
3. + 250 to +400%
4. Over +400%
E. Over +100%
F & G. + 45 to +70%
H1 & I1. −30 to −45%
H2 & I2. −45 to −75%

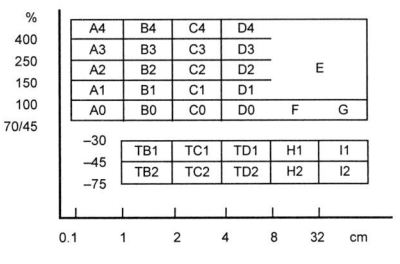

Very short thick fault (A0, B0, C0 and D0)
Short thin faults (TB1, TC1, TD1, TB2, TC2, TD)

Figure 4.32: Different size and length groups of classified faults

Faults are cumulative i.e.

$$A_1 \text{ Faults} = \text{sum of faults of } A_1 + A_2 + A_3 + A_4,$$
$$A2 \text{ Faults} = \text{sum of faults of } A_2 + A_3 + A_4,$$
$$A3 \text{ Faults} = \text{sum of faults of } A_3 + A_4,$$
$$A4 \text{ Faults} = \text{Only } A_4 \text{ faults.}$$

Likewise: $B_1 \, B_2 \, B_3 \, B_4 / C_1 \, C_2 \, C_3 \, C_4 / D_1 \, D_2 \, D_3 \, D_4$,

H_2 Faults = sum of the faults of $H_1 + H_2$,

I_2 Faults = sum of the faults of $I_1 + I_2$.

The faults are always expressed as ACTUAL faults and not cumulative. An example will make this point clear.

20s Carded yarns were tested and printout shown per 100 km is as given in Table 4.7:

Table 4.7: Fault classification

Class	Length (cm)	Cross-section	Cumulative	Actual
A_1	0.1–1	+100% to +150%	76.0	66.6
A_2	0.1–1	+150% to +250%	9.4	1.2
A_3	0.1–1	+250% to +400%	8.2	8.2
A_4	0.1–1	+400% and above	0.0	0.0

A_1 (Actual) = 76 − 9.4 = 66.6.

A_2 (Actual) = 9.4 − 8.2 = 1.2.

A_3 (Actual) = 8.2 − 0 = 8.2.

A_4 (Actual) = 0.0.

Out of the above 23 number of faults, the presence of certain faults if present in the outgoing yarns will give rise to the working problems in forward integration and also impair the quality and appearance of the fabric. These faults are to be removed while in winding stage and minimum permissible presence of such faults is allowed to pass depending on the choice of customer. These faults are:

1. Objectionable faults/100 km = $A_4 + B_4 + C_4 + D_4 + C_3 + D_3$.
2. Maximum permissible limit to be present = faults 0.8 − 1.0 per 100 km.
3. Serious faults/100 km = $A_3 + B_3 + C_2 + D_2$ = 3–4 faults/100 km.

4. Long thick faults/100 km = E + G = 1–2 faults/100 km.

5. Long thin faults/100 km = I_1 + I_2 + 5–6 faults/100 km.

The classimat print out shows the yarn's faults in two colours, one in black and other in red, in the same sheet. Black coloured shows the actual length of yarn fed for testing. Red coloured shows the faults converted to 100 km. Just one example will make it clear, suppose A1 faults in black ink shows 500 faults for feeding 120 km of yarn, red ink print out will show it for 100 km i.e. 500 × 100/120 = 417 faults. Likewise, all faults will be converted to 100 km length of yarns.

The short thick faults (A1–D4) of classimat system are either caused by the raw material defects and preparatory stages or due to the drafting defects. If a diagonal is drawn joining A4 and D1 then the matrix is divided into two equal parts as shown in Fig. 4.33. According to the thumb rule of spinning, the faults lying within the upper triangle are due to the drafting faults whereas the faults lying below the diagonal are either due to the deficiency in the raw material or due to the opening problem of the blowroom and carding machines.

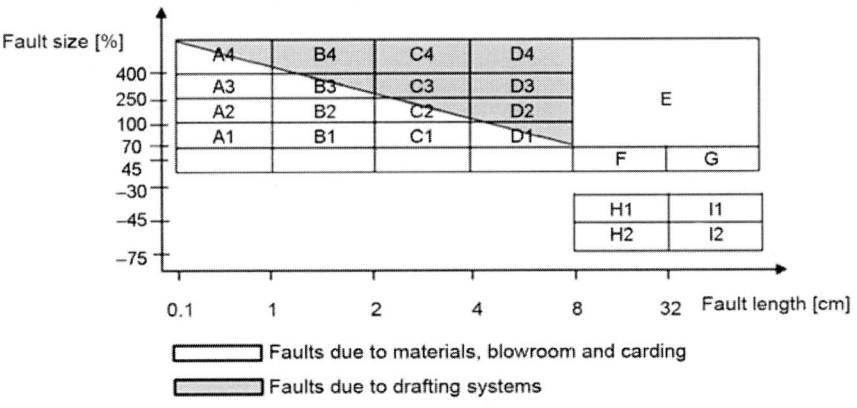

Figure 4.33: Thumb rule to locate source of faults

Uster® classimat quantum

Today with the Uster® classimat quantum generation, the cleared and uncleared yarn can be checked; the classification of thick and thin places, neps and foreign fibres in the yarn can be fulfilled. This system also helps the user in determining the optimal limits for yarn clearing, in analysing new material and supports with experience values, which can be used for benchmarking and evaluation.

With Uster® classimat quantum, there is also a possibility of defining new customised classes called "Tailored classes". The system can classify and present the events in the tailored classes for thick places, thin places and foreign fibres.

The scatter plot is a very important feature of Uster® classimat quantum which helps the user in analysing the exact place of each event in the classification matrix and indicates the yarn faults of both the standard classes and the extended classes as points in the classification matrix. The exact length and cross-section increase of the individual yarn faults can easily be determined with the horizontal and vertical scales.

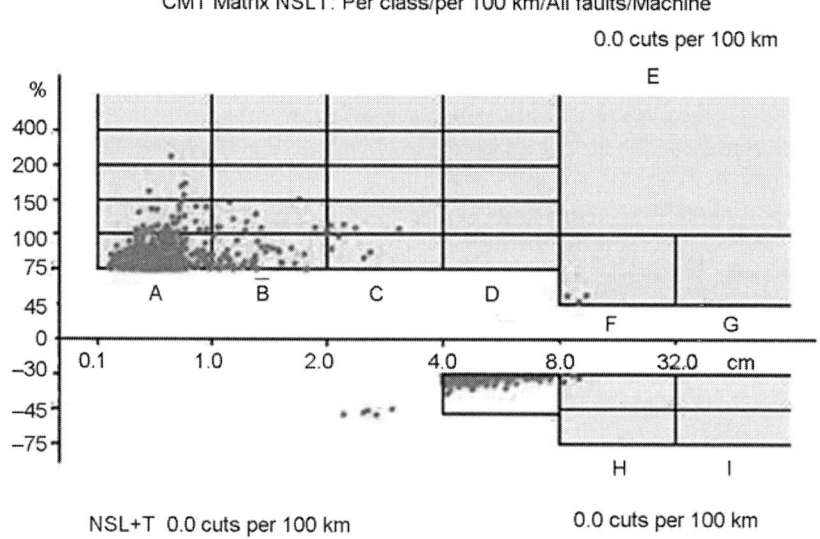

Figure 4.34: Uster® classimat quantum matrix

Classification matrix for foreign fibres

In addition to the standard classification, this system also allows the user to measure foreign fibres and vegetables in a yarn and classify these faults in 27 foreign fibre classes, as shown in Fig. 4.35. No classification data is available for the A1 class, because there are no significant faults in this class. With the vegetable filter, it is possible to differentiate between organic and synthetic foreign fibres. Based on the fact that vegetables mostly do not have a disturbing effect on the appearance of fabrics, because they can be bleached or can absorb the same dyestuff, these particles are allowed to remain in the yarn for many fabrics and, as a result, it saves a considerable number of cuts on the winding machine and reduces the formation of splices.

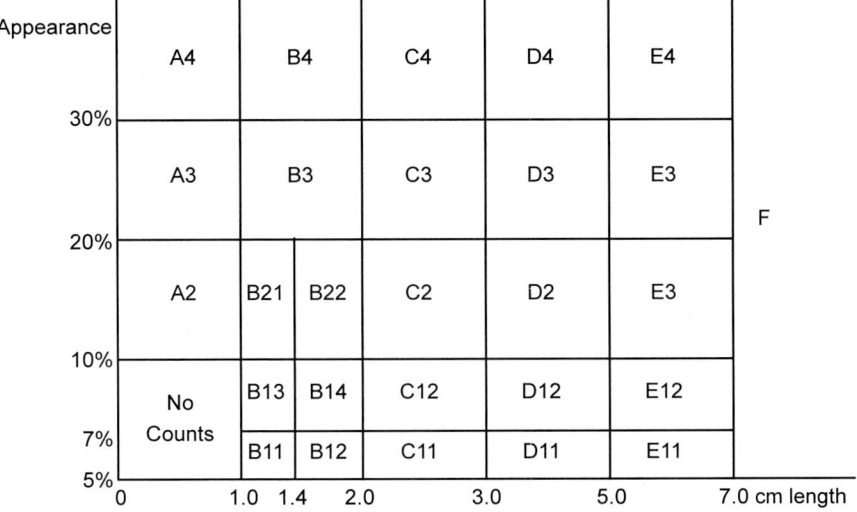

Figure 4.35: Classification matrix for foreign fibres

4.2.9.1.2 Optical method

The principal of measurement of yarn fault using this method is as follows:

The very familiar instruments available using this principal of measurement is "Classifault (CFT-II)" developed by Keisokki Kogyo Co. Ltd. Photo electric sensor is used for measuring faults and the output is given in terms of cross sectional size and length. The classifault system is shown in Fig. 4.36. This system gives 40 classification channels for grading yarn faults. In addition, it has software flexibility for changing the limit level of classification and can record the fault in the form of histogram.

The CFT-II classifies the yarn faults in 40 classes – slubs into 20 classes, thick place into 10 classes and thin place into 10 classes and provides the measured results for winding position as well as total number of them via the printer. Obtained results as well as test condition can be stored in the disk allowing latter analysis for extracting only required data of particular type of faults.

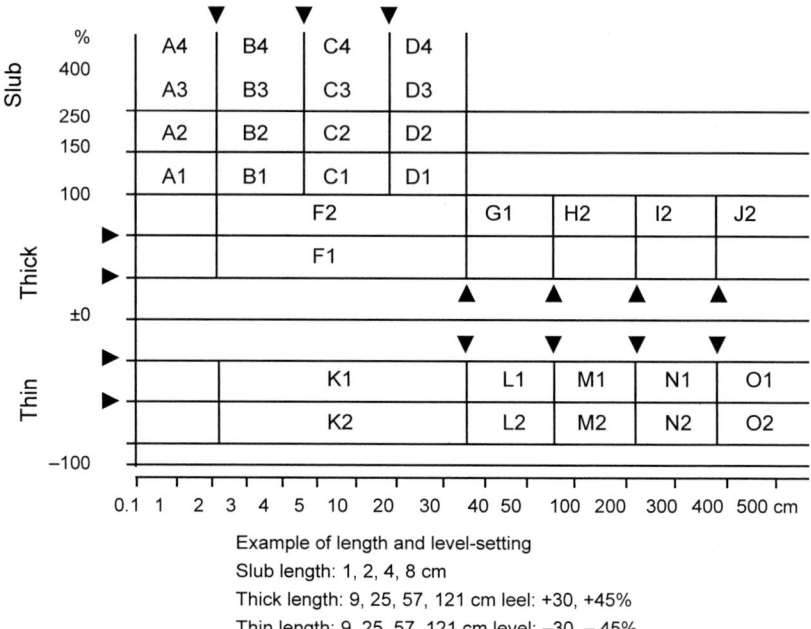

Figure 4.36: Classification matrix for classifault (CFT-II)

1. Classification of yarn faults

Various literatures have classified yarn faults differently. Thomson [93] consolidated the 16 classes of faults obtained from classimat to sources, and they are shown in Table 4.8:

Table 4.8: Type of yarn faults

Type of yarn faults	Classimat classes
Drafting faults	C1–C2–C3–C4, D1–D2–D3–D4
Waste slubs	A1–A2–A3–A4, B1–B2–B3–B4
Fly slubs	A3–A4, B1–B2–B3–B4, C2–C3–C4, D3, D4
Operating faults	B3–B4, C2–C3–C4,D1–D2–D3–D4
Contact faults	A1–A2–A3–A4, B1–B2–B3–B4, C2–C3, D3–D4
Fibre on blend faults	A1–A2–A3, B1–B2, C1, D1–D2

Hattenschariffer and Nabnit [94] have classified faults on the basis of their source of generation i.e. whether they are due to raw material, process or spinning related. Table 4.9 gives the comprehensive classification and the finding of the study done in terms of the contribution of each category to the total fault.

Table 4.9: Classified faults on the basis of their source of generation

Short staple cotton spinning system	Carded cotton in %	Combed cotton in %	Blend yarn of manmade fibre and cotton in %	100% Manmade fibre in %
M 1 Foreign matter	16.2	11.1	6.4	1.9
M 2 Fibre entanglements	–	4.8	10.0	23.0
M 3 Synthetic undrawn fibres	–	–	1.3	5.3
P 1 Piecing	10.1	12.5	8.6	16.0
P 2 Long slub	–	–	–	–
P 3 Short slub				
S 1 Spun-in fly	42.0	44.0	38.0	29.0
S 2 Loose fly	18.0	14.3	22.0	15.0
S 3 Long collections of fly	3.0	1.0	4.3	2.0
S 4 Fishes(corkscrew type faults)	–	0.8	1.3	1.3
S 5 Pushed-together collections of fly	2.0	3.1	1.0	0.1
S 6 Chains of faults	2.1	1.3	1.3	0.3
S 7 Crackers	0.4	0.2	0.4	–
S 8 Over-twisting	–	–	–	–
Size of sample	2834	3528	2466	1382
Number of faults checked	276	243	153	115
Number of qualities	137	118	105	89
Number of mills				

Short description of the faults:

• M1 – foreign matter

Due to non-textiles material which is already available in the bales or is collected at some stage during the spinning process.

• M2 – fibre entanglements

These entanglements are found primarily in yarn containing man-made fibres. They consist of fibres which are bound together and in many cases, are combined with collection of finish material. In groups they grow to become thick faults.

• **M3 – synthetic undrawn fibres**

These are, of course, only encountered in man-made fibre yarn and are the results of yarn and are the results of single fibre stuck together and of particles of synthetic material.

• **P1 – piecing**

These piecings are normally produced during the processes prior to spinning.

• **P3 – short slubs**

These are due primarily to collection of short fibre which are not drafted by the roller drafting system and appear as thick places. They contain little twist and are, accordingly weak in strength. This type of faults can also be the result of too wide setting of the gauge with apron drafting.

• **S1 – Spun-in fly**

This refers to free fibres which fall in to the drafting elements or on to the roving fed into the drafting unit, and are then twisted in to the yarn along their entire length.

• **S2 – loose fly**

This refers to fibres which are collected by the yarn at a position after the front roller and, in one end.

• **S3 – long collection of fly**

These are matted fibres which, collect together on aprons or rollers and from time to time are collected and carried along by the yarn.

• **S4 – fishes (corkscrew type faults)**

These faults are due to static charging of fibres as a result of unsuitable drafting aprons or drafting aprons which have cracked surfaces.

• **S5 – pushed together collections of fly**

These are faults resulting from held back fibres, and occur primarily at the ring traveller.

• **S6 – chains of faults**

These are combination of the faults S1, S2 and possibly also S3 which occur in short succession, are after the other, along the length of the yarn.

• **S7 – crackers**

These results due to extra-long fibres which disturb the drafting process and, for a short instant of time, stop the passage of the yarn.

2. Sources of yarn faults and factors contributing to their generation

Sources of yarn faults:

The information available on this is subdivided into two categories:

(a)   Class wise

(b)   Type wise

Class wise:

Vijayshankar and Gupta [95] mentioned in their paper that:

'A' faults – 80–90% of the 'A' faults contain a nucleus of seed coats fragments with adhering fibres that bind the fragment to the body of yarn.

'B' fault – In the case of B faults, seed coats fragments account for 30–60% of the total 'B' faults. Seed coats fragments account for most of the 'A' faults but only for about 50% of the B faults. In other words, apart from seed coats fragments some other factors contribute significantly to 'B' faults.

'C and D' faults – Factors like spun in lint, loose lint, bad piecing, etc. are known to contribute to 'C and D' faults. Drafting deficiencies are not likely to have a very significant contribution to the frequency of 'C and D' faults in the yarn, as these faults are recorded at a level of yarn weight per unit length +100% or more. Unopened fibre cluster that are present in the sliver and are potential slubs in yarn, have the largest processing contribution for 'C and D' types of faults.

Garde [96] in his paper has clearly mentioned that the Indian polyester/cotton yarns have very high incidence of each type of faults. The faults of 'A' type are almost six times as numerous, while those of other types are two to four times more than the Uster statistics. The most disturbing feature is that even the objectionable faults are four times more.

Kumarswamy and Sheriff [97] conducted a study with count group from 29s Ne to 40sNe over the assessment of yarn faults and observed that in Indian mills, 'A' type of faults are 10–15 time higher and 'B' and 'C' types of faults are 5–6 times higher in comparison to international standard of respective type faults.

Ramchandran [98] mentioned in his paper that A4 and B4 faults come due to fly at ring traveller; C3, C4, D3, D4 faults come due to defective drafting elements at roving; B, C faults due to excessive trash and I1 due to roving stretch.

According to Yadav [99] H1 faults increase in winding after spinning. Old apron sets cause more H1 and any hindrance to cause tension in roving fed results more H1. H1 is lower with higher spindle speed and lower break draft. Variation in fibre length also causes reduction of H1. Mechanical or pneumatically arm loading has no significance in H1 faults and also higher draft of spinning is prone to higher number of H1 faults.

Type wise:

Hattenschwiler and Bubler [100] mentioned in his paper that the shorter and smaller size faults are due to the raw material, the opening and the carding processes; the longer and larger size faults are introduced in the processes just prior to spinning and during spinning.

It has been generalised that quarter of all faults are due to the raw material and process prior to spinning and approximately three quarters are introduced in to the yarn at the spinning machine.

A basic rule in quality management is a preventive maintenance rather than corrections afterwards. Unfortunately, this is not yet possible with the technology of today. Textile specialists in spinning mills who have to conquer disturbing yarn fault have to find the origin of such yarn faults.

Table 4.10 shows a selection of sources, which produced seldom-occurring faults in the respective categories. It is a collection of reasons over many years why such events happened [101].

Table 4.10: Selection of sources producing seldom-occurring faults

Classes		Possible reason of faults	Comments
A (Thick place)	A0	Extended class, mainly used for ply yarn	
	A1	Bad condition of carding, blow room, trash in yarn	(Short thick place)
	A2	Bad condition of carding, blow room, trash in yarn	
	A3	Neps, fluff, foreign matters, dirty drafting zone	
	A4	Ring front zone dirty, fly waste in trumpet	(Unacceptable faults)
B (Thick place)	B0	Extended class, mainly used for ply yarn	
	B1	Fibres damage in process, spindle without aprons	(Short thick place)
	B2	Fibres damage in process, spindle without aprons	
	B3	Fluff in travellers, unsuitable travellers, bad piecing	
	B4	Slub from ring spinning department	(Unacceptable faults)
C(Thick place)	C0	Extended class, mainly used for ply yarn	
	C1	Bad piecing in cans, sliver entanglements	(Short thick place)
	C2	Bad piecing in cans, sliver entanglements	
	C3	Piecing, ring spinning	(Unacceptable faults)
	C4	Floating fibres, fly, slub	(Unacceptable faults)

Contd...

Contd...

Classes		Possible reason of faults	Comments
D (Thick place)	D0	Extended class, mainly used for ply yarn	
	D1	Floating fibres	
	D2	Gauge problem of roving frame, spacer problem	(Unacceptable faults)
	D3	Fluff in ring spinning department	(Unacceptable faults)
	D4	Fluff in ring spinning department	(Unacceptable faults)
E (Thick place)	E	Double yarn	(Spinners double)
F (Thick place)	F	Bad piecing in ring yarns, roving & back process	(Long thick places)
G (Thick place)	G	Bad piecing in ring yarns, roving & back process	(Long thick places)
H (Thin place)	H1	Mostly eccentric bobbins in roving &ring frames, eccentric spindles	(Thin places)
	H2	Poor handling of materials during processes	(Thin places)
I (Thin place)	I1	This type of faults is produced by separation of parts of sliver or roving prior to spinning	(Long thin places)
	I2	This type of faults is produced by separation of parts of sliver or roving prior to spinning	(Long thin places)

4.2.9.2   *Factor contributing to the high incidence of the classimat fault*

Pillay and Ratnam [102] in their paper have drawn the following conclusions that,

(a) Faults due to raw material are about 8 times greater in cotton than in staple fibres whereas the drafting faults are only 1.5 times greater. The numbers of objectionable faults are also 1.5 times higher in cotton than in staple fibre yarn, and staple fibre yarn give these C and D type of faults when compared to cotton.

(b) The number of faults systematically increases with increase in count both in carded and combed yarn however percentage increase is lower for the latter than the former.

(c) Use of high production and tandem cards as compared with semi-high production card as well as flexible fillet carded yarn, significantly reduce yarn faults. Higher percentage of waste removed in carding and combing lowered the fault level appreciably.

(d) According to them [102] the total number at faults in combed yarns is about 65–85% lower than those in corresponding carded yarn.

(e) As the percentage of noil removed increased, objectionable fault reduced considerably.

(f) Conventional drawing gives lowest number of faults per 100 km, and semi high production andhigh production drawing give substantial increase in the incidence of faults.

(g) Overhead clearer generally used by mills to keep the department clean and prevent fly and dust depositing on the yarns help in reducing the incidence of different type of yarn faults. In general, it is observed that total faults in cotton yarn are reduced by about 50% and in staple yarn by 50%.

(h) No firm conclusion could be drawn regarding the effect of drafting system and spindle speeds on yarn faults.

Kumarswamy and Sharieff [103] mentioned in their paper that high production and semi-high production card do not have any effect on faults. Carding rate affect C and D type of faults which increase marginally with carding rate. Objectionable faults A4 and B4 faults are affected by carding and objectionable faults C3, C4, D3 and D4 are affected by the ring frame only. Metallic and tandem card reduce A1, A2 andA3 types of faults significantly. The long faults are affected marginally and objectionable faults remain unchanged. It appears that tandem card reduce only the short length faults.

Vishwanath et al. [104] mentioned in their paper that about 60–70% of objectionable faults are due to ring frame while 30–40% of them are due to preparatory. They found that ring frame equipped with pneumatic loaded drafting system gives all round best results.

According to Bhatt et al. [105], if the clearing procedure is not performed satisfactorily by the ring frame top roller clearer than A, B1, B2, C1 andC2 type faults will be higher.

According to them [100] piecing shares 9–16% of the total number of disturbing yarn faults. The consistency of their number depends mainly on the piecing at pre-spinning process.

Short slubs occur to an extent of 5–7% in all yarns. The drafting systems at various processes, short fibres and humidity in the department are important in its formation. The spun-in-fly type faults may be as high as 50% or more of all objectionable faults in any yarns. Therefore, housekeeping and humidity in the department are very important. Objectionable faults introduced at spinning process account for 70% in case of carded yarn and 50% in case of man-made staple fibres yarns.

They concluded from their study that the total yarn faults in Indian yarn are 9–12 times the international standards whereas objectionable faults are to an extent of 5–7 times. Ring frame is the highest contributor to the objectionable faults and drafting systems, heart of ring frame, is the culprit of faults generation at the same time.

The overhead blower reduces 12.5% of all faults. The reduction in C, D faults is 27.3% and that for objectionable fault is 28.3%. That is, such a high number of faults are influenced by the possibility of blowing (cleaning) or picking-up of fly.

Another observation is that most of the mills are not maintaining controlled atmosphere, the department temperature and humidity changes from time to time in a day or day after day.

4.2.9.3   *How to control long thin faults in yarns*

Following points if taken care of, will certainly reduce long thin faults in the yarns:

(i)   Improper sliver can maintenance causes long thin places to increase. Inadequate spring tension, damaged edges, damaged rivets inside can, unbalanced spring, etc. should be mended.

(ii)   Adequate gap (2–3 mm) between sliver coil and can wall is to be maintained throughout the circumference of the can.

(iii)   Slivers should never be touched by finger.

(iv)   Proper adjustment of tension draft in drawing machine (creel and web) and simplex (creel) is to be maintained.

(v)   High roving stretch due to low twist in roving or wrong rachet wheel adjustment is to be taken care of.

(vi)   False twist effectiveness is to be checked from time to time through 60 yd roving lea strength check.

(vii)   Broken simplex bobbin holder pins are to be mended.

(viii)   Proper arrangement of drawing cans in simplex creel needed.

(ix)   Creel roller of simplex machines should be in rotating condition.

(x)   Drawing can's sliver should be taken over creel roll not over creel rod.

(xi)   Jammed umbrella creel in ring frame is either to be attended or discarded.

(xii)   Criss-cross roving arrangement in ring frame creel is never to be done.

(xiii)   Creel rod in ring frame should be free from fluffs and flies and creel rod should be properly polished.

4.2.9.4   *Clearing of faults*

Regarding the clearance of faults from the yarn through eye, short thick faults and long thick faults are detected by EYC and they are eliminated, but long thin faults like H and I are seldom cut (EYC like UPM-200 and UPC-200 are taken into consideration here). Special care is to be taken in spinning and spinning preparatory process for controlling long thin faults. Also one good habit of auto winder if practiced gives a very good control for eliminating finer count bobbin and hence long thin places.

When a new or fresh bobbin is creeled in Auto corner (here Savio Espero L fitted with UPM-200 EYC is considered), then it is first taken on count alarms channel and then goes to general channel. If a new bobbin is rejected consecutively three times after creeling the computer setting should be such that red light lit up and the drum will stop. The operator of the Savio should take out 2–3 g of yarns from the rejected bobbin and corresponding cone. The operator should again creel that bobbin and if again the bobbin is rejected, the operator should keep the bobbin in a basket marked as rejected bobbins. QA department should collect all such bobbins shift wise and take the wrapping of all such bobbin shift wise to find out the trend of wrapping in this bobbin. This practice helps to control thin places in the cones and also helps to take corrective action in back processes. Our aims should be to control faults from the back process and not to eliminate or cut the faults in auto corner. Excess cuts and corresponding splices in the cones increase the chances of splice slips in the forward process.

4.3   References

1. Hsieh, Y.L., 1999, Structural Development of Cotton Fibers and Linkage to Fiber Quality, p. 137–165. In A.S. Basra (ed.) Cotton Fibers. The Haworth Press, Inc., Binghamton, NY.

2. Ramey, H.H., Jr. 1999. Classing of Fiber. p. 709–727. In C.W. Smith and J.T. Cothren (eds.) Cotton: Origin, History, Technology, Production. John Wiley & Sons, Inc., New York, NY.

3. Bragg, C.K. and F.M. Shofner, 1993. A rapid, direct measurement of short fiber content. Textile Res. J. 63, 171–176.

4. Shofner, F.M., Williams, G.F., Bragg, C.K.and Sasser, P.E., 1988, Advanced fiber information system: A new technology for evaluating cotton, Conference of the Textile Institute, 7–8 Dec.. Coventry, UK.

5. Shofner, F.M., Chu, Y.T.and Thibodeaux, D.P., 1990, An overview of the advanced fiber information system. p. 173–181. In Proceedings of International Cotton Conference, Faserinstitut, Bremen, Germany.

6. Mor, U., 1996, FCT—Fiber Contamination Tester—a new instrument for the rapid measurement of stickiness, neps, seed-coat fragments and trash—for the ginner to the spinner. In H. Harig, S.A. Heap, and J.C. Stevens (eds.) 23rdInternational Cotton Conference, Proceedings International Cotton Conference, Bremen, March 6–9, 1996, pp. 205–212. Faserinstitut, Bremen, Germany.

7. Lintronics. 2000. Standard Test Methods for Cotton Fiber Amount of Stickiness and Determination of its Severity. In S.A. Heap and J.C. Stevens, Proceedings International Committee on Cotton Testing Methods, Bremen, February 29-March 1, 2000, appendix S-2. International Textile Manufacturers Federation, Zurich, CD-ROM.

8. Slater, K., 1986. Yarn evenness, Text.Prog.,14, No. 3/4.

9. Martindale, J.G., 1945. A new method of measuring the irregularity of yarns, J. Text. Inst. 36, T 35.

10. Foster, G.A.R., 1958. Manual of Cotton Spinning,Vol. IV, Pt I, Textile Institute and Butterworths, Manchester, London.

11. Sitrafocus,Vol.5, No 1, May 1987.

12. Basu, A., 2001, Textile Testing: Fiber, Yarn and Fabric, SITRA, Coimbatore, pp. 211–226.

13. Catling, H., 1958, Some effects of sinusoidal periodic yarn thickness variations on the appearance of woven cloth. J. Text. Inst. 49,T232–T246.

14. Lunenschloss, J. and Helli, J.G., 1971, Investigation of the Thickness Frequency of Different Cotton Assortments over a Longer Production Period. Z. ges. Textil-Industrie 73, 657.

15. Czaplicki, Z., 1973, Effect of the secondary wall of cotton fibre on the number of faults in a web and on the irregularity of yarn spun with the ring and open-end systems.Przeglad. Wlok.27,178.

16. Seshan, K.N.,et al. 1979. Cotton in a Competitive World. p. 202.In P.W. Harrison (ed.), The Textile Institute, Manchester.

17. Ratnam, T.V., Seshan, K.N. and Govindarajulu, K., 1974, Some factor affecting yarn irregularity. J. Text. Inst., 65,61.

18. Characterization and Quantification of Woven Fabric Irregularities using 2-D Anisotropy Measures. (Under the direction of Dr. Warren J. Jasper and Dr. Moon W. Suh.), North Carolina State University, July2005.

19. Eric Oxtoby; Spun Yarn Technology, Butterworth & Co (Publishers) Ltd., 1987, p. 51.

20. Physical Testing of Textile, B P Saville,Textile institute, Cambridge, England, 2000.

21. Hearle, J.W.S., P. Grosberg, and S. Backer. 1969. Structural Mechanics of Fibers, Yarns, and Fabrics. Wiley-Interscience, New York, NY.

22. Alberto Barella, 1952, The influence of twist on the regularity of the apparent diameter of worsted yarns. J. Text. Inst. 43, P734–P741.

23. Uster Technologies AG, Switzerland. Uster Tester 4-SX, 2004.

24. Pillay, K.P.R.,and R. Hariharan. Sept 1984. Effect of Processing Factors on the Incidence of Yarn Faults in Spinning. Ind. J. Text. Res. 9, 100–105.

25. Unwin, H., and J.W. Reast. 1950, The Effect of Yam Irregularities on Fine Gauge Fashioned Hosiery,.J. Text. Inst. 41, P547.

26. McFarlane, R.A. 1950, Yarn irregularities and their effect on viscose rayon fabrics, J. Text. Inst. 41, P566.

27. Bowles, A.H. 1972, Control, Shirley Institute, Manchester.

28. Piso, J. 1975, Capacitive measuring system, Mod. Text. 56, No. 3, 46.

29. Snowden, D.C., and J. Sidi. 1950, The efficient mixing of irregular weft yarns, .J. Text. Inst. 41, P507.

30. Bleakley, T. 1950, The importance of irregularity of linen yarns ,J. Text. Inst.41,P526.

31. Zellweger Ltd. 1971.Uster News Bull.Jan., No. 15.

32. Mahajan, S.D. 1971, Periodic yarn count variation & weft bars, Text. Dig. 32,145.

33. Garde, A.R., S. Bandyopadhyay, and T.A. Subramanian. 1972. Influence of yarn unevenness and cover factor on fabric appearance, Proc.13th Tech. Conf. ATIRA, BTRA,and SITRA.1, 12.

34. Warty, S.S., and T.V. Talele. 1974. Problems in wet-processing due to irregularity of yarn, Man-made Text. India. 17, 349.

35. Louis, G.L., Grading open-end spun cotton yarns, 1978.American Text. Rep./Bull. Edn.AT7, No. 4, 88.

36. Barker, S.G., A gravimetric method for investigation of the variation and levelness of yarn, 1926.J. Text. Inst. 17, T259.

37. Martindale, J.G., A new method of measuring the irregularity of yarns with some observations on the origin of irregularities in worsted slivers and yarns, 1945.J. Text. Inst. 36, T35.

38. Harry B. Iler Jr., An Orderly Warp Storage Room Saves Time and Effort , Text. World,1952, 102, No. 12, 136.

39. Frenzel, W., 1922, Determination of the yarn thickness with a measuring apparatus,. Leipzig. Monats. Text. Industr. 37, 166.

40. Oxley, A.E. 1922, The Regularity of Single Yarns and Its Relation to Tensile Strength and Twist, J. Text. Inst. 13, T54.

41. Anderson, Cavaney, Foster, and Wormersley., An automatic sliver and roving regularity tester and an automatic yarn regularity tester, 1945. J.Text. Inst. 36, T 253.

42. Anderson, Cavaney, Foster, and Wormersley., Description and use of (1) The photographic yarn regularity tester, and (2) The photographic sliver and roving regularity tester, 1945. J.Text. Inst. 36, T 191.

43. Instructions of Koth evenness tester.

44. Catalogue of "Fielden-Walker" yarn evenness tester.

45. Fielden. 1946. Dielectric Comparator, B. P. 619, 534.

46. Hasler, A and Honegger, E., Yarn evenness and its determination, Text Res J 1954; 24: 73–85.

47. Methods of measuring yarn unevenness, JSIF (Japan Spinners Inspection Foundation). 1952. No. 3, 20.

48. Kato, Sakaoku, and Yoshida. 1952., Yarn thickness measurement by electric resistance,Rep. Fau. Eng. Fukui Univ., No. 1, p. 17.

49. Nozaki, Aino, Ando, and Hasegawa., 1953, Electric resistance strain gauge, Rep. Gov. Ind. Res. Inst., Japan, 2, 1.

50. Operating instructions for Uster Integrator ITG1.

51. Oxley, A.E., The Regularity of Single Yarns and Its Relation to Tensile Strength and Twist, 1922. J. Text. Inst., 13, T 54.

52. Keinath, G., The Measurement of Thickness, p24, NBS Handbook, Issue 77, Volume 3, U.S. Government Printing Office, December 1958.

53. Saxl, I.J., An Evenness Tester, 1935. J. Text. Inst., 26, T 77-82.

54. Saco-Lowel Tester, 1945, Text Recorder, 63, 38.

55. Martindale, J. G., 1941. A correlation periodograph for the measurement of periods in disturbed wave-forms, J. Text. Inst, 32, T 71

56. Stanbury, G.R., 1932,The measurement of the levelness of worsted yarns., J. Text. Inst., 23, P197-206.

57. Strother, F.P., 1952, Photoelectric device, filometer to measure diameters of threads, Electronics, 25, 110. .

58. Starr, A.T., 1932, A Note on Impedance Measurement, Wireless Eng. Exp. Wireless, 9,615.

59. Mathew, J. A., Raichenbaum, N. B. & Spencer-Smith, J. L, 1950. A review of methods of measuring the irregularity of flax roves and yarns, J. Text. Inst, 41, P486

60. Walker, P.H., 1950, The electronic measurement of sliver, roving and yarn irregularity, with special reference to the use of thr fielden bridge circuit, J. Text. Inst. 41, P446.

61. Sust, A. and A. Barella. 1964. Twist, diameter, and unevenness of yarns a new approach, J.Text. Inst. 55, T1.

62. Spencer-Smith, J.L., and H.A.C. Todd. 1941. A time series met with in Textile Research, Suppl. J. Rov. Stat. Soc., 7, 131.

63. Chamberlain, N.H., 1944, A general-purpose photoelectric photometer and its use in textile laboratories, J. Text. Inst. 35, T61.

64. Martindale, J.G., 1945, A new method of measuring the irregularity of yarns with some observations on the origin of irregularities in worsted slivers and yarns, J. Text. Inst. 36, T35.

65. "Uster", Electrical Sliver, Roving and Yarn Regularity Tester. U.S.A. Uster Corporation. Textile World, 1948, 98, No. 6, 140.

66. Cox, J. D.R. and Townsend, M.W.H., 1951, The use of correlograms for measuring yarn irregularity, Text. Inst. 42, P145.

67. Picard, H.C. 1951. The irregularity of slivers-I , J. Text. Inst. 42, T503.

68. Picard, H.C. 1952. The irregularity of slivers-II , J. Text. Inst. 43, T251.

69. Van Zwet, C.J. and Cox. D. R, 1955. A method for the calculation of the cb (l) curve. J. Text. Inst. 46, P794.

70. Hoffmann, D. 1955.Coefficient of variation quantifying irregularity, Industr. Text. 821.

71. Grosberg, P. 1956., "Letters to the editor", J. Text. Inst. 47, TI79.

72. Dyson, E. 1973., Some observations on yarn irregularity, J. Text. Inst. 64, 215.

73. Bowles, A.H., and I. Davies., Yarn quality - The Influence of Drawing and Doubling Processes on the Evenness of Spun Yarns, Text. Inst. Industr., Volume 16, Number 1, January 1978.

74. Zellweger Ltd. 1962.Uster News Bull. June, No. 2.

75. Spencer-Smith, J.L., and H.A.C. Todd. 1941, A time series met with in textile research, Suppl. J. Roy. Stat. Soc. 7, 131.

76. Martindale, J.G. 1945. A new method of measuring the irregularity of yarns with some observations on the origin of irregularities in worsted slivers and yarns, J. Text. Inst. 36, T35.

77. Huberty, A. 1947. First study of the parameters characterizing the regularity of the threads, wicks and ribbons- Fundamental laws, Proc. IWTO Tech. Cttee.1, 55.

78. Furter, R. 1982., Evenness Testing in Yam Production, Part I. The Textile Institute, Manchester.

79. Lund, G.V. 1952., A comparative study of the regularity of rayon staple yarns spun on different spinning systems J. Text. Inst. 43, T299.

80. Grosberg, P., and R.C. Palmer.1954., The use of the Zellweger irregularity tester in finding the variance-length curve of worsted yarn, J. Text. Inst. 45, T273.

81. Grosberg, P., and R.C. Palmer., 1954., On The Determination of the $B–L$ Curve by Cutting and Weighing, J. Text. Inst. 45, T291.

82. Van Zwet, C.J. 1955., A Method for the Calculation of the CB(L) Curve, J. Text Inst. 46, P794.

83. Nienhuis, W.A. 1963., A simple method for determining the CB(L) curve by cutting and weighing, J. Text.Inst. 54, T353.

84. Dyson, E., and B. Schofield. 1968, A new method for determining the variance-length curve of yarns and slivers, J. Text. Inst. 59, T528.

85. Nute, M.E., W.R. Pelton, and K. Slater. 1972, The variance between ultra-short lengths of yarn , J. Text. Inst. 63, 212.

86. O'Connell, R.A.,F.J. Ahrens, and R.J. Martsch. 1975, Unevenness (Length Variance) of a Worsted Yarn by Cutting-and- Weighing and the Use of the Pacific Tester, Text. Res. J. 45, 596.

87. Tippett, L.H.C. 1935. Some applications of statistical methods to the study of variation of quality in the production of cotton yarn. Suppl. J. Roy. Stat. Soc.2,27.

88. Martindale, J.G. 1943. Irregularity in worsted rovings and yams. W.I.R.A. Bull. 9.

89. Townsend, M.W.H., and D.R. Cox. 1951, The analysis of yarn irregularity, J. Text. Inst. 42, P107.

90. Principle of Textile testing by J.E. Booth, Chapter9, Page 466.

91. de Malatinszky, P.,and P. Grosberg. 1974, The Medium- and Long-Term Variations of a Yarn—II, J. Text. Inst., 55, 46, 310.

92. Locher, H., and H. Ernst, Quality Control and Supervision of Yarn Faults in the Spinning Mill., Uster News Bulletin, No. 17.

93. Thomson, W.A., A New Era in Quality Control – Yarn Fault Management, Uster News Bulletin., No. 18.

94. Hattenscharitter, P., and B. Nabnit, The source and frequency of yarn faults, Uster News Bulletin., No. 21.

95. Vijayshankar, M.N., and A.K. Gupta, Origin and Control of Faults in Cotton Yams, ATIRA., Oct. 1984, Page 1.

96. Garde, A.R., 37th All India Textile Conference., 1998, Page 87.

97. Kumarswamy, K., and Sheriff, I. 20th Joint Technological Conference of A.B.S., 1997, Page 7.1.

98. Ramachandran, V., Effective utilisation of yarn fault information to gain quality, Sept–Oct 1996. J. Text. Assoc. (Bombay). No 3, P 153.

99. Yadav, R.N., May 1993, H1-Uster classimat fault – experience, Manmade Textile India, 36, No. 5, Page 185–188.

100. Hattenschwiler, P., and M.Bubler, Uster News Bulletin No 21, Nov. 1973.

101. DÖnmez Kretzschmar, S.,andR. Futer, Uster Classimat Quantum,Application Report, May2008.

102. Pillay, K.P., and Ratan, T.V., Dec. 1981, An inter firm comparison of yarn faults by Uster Classimat system, SITRA Vol. 21, No. 3.

103. Kumarswamy, K., and Sharieff, Nov. 1979. Infrequent yarn faults: their incidence, causes and removal, J. Text. Assoc. 40, 6, Page 213.

104. Viswanath, C.S., Jumdar, C.R., et al., Oct. 1991, Ring frame drafting system contributing to yarn faults? ITJ, 102, No. 1, 90–93.

105. Bhat, P., Kane,C.D. et al., Apr. 1993, Kane, Influence of ring frame top roller cleaning method on classimat faults,Technological Conference, Resume of Papers, BITRA, SITRA, NITRA and ATIRA, pp. 65–72.

5

Application of statistical tools in day to day work

There are some important statistical tools which are used to check the reliability of the testing results and observations (test of significance). These are discussed below:

There are two types of variables which cover the yarn parameters and important observations in spinning process. They are: (i) Continuous variables and (ii) discontinuous variables.

(i) Continuous variables come under the normal distribution and discontinuous variables come under binomial or Poisson's distribution.

(ii) Continuous variables may occur in fraction but discontinuous variable can never come in fraction.

Continuous variables	Discontinuous variables
Count	Neps
Lea strength	No. of laps
Twist	End break
Single yarn strength	Faults/100 m
Lap rejection%	
Lap weight	
Waste%	
Efficiency	
Hairiness	

5.1   Common statistical tools used for continuous variables

Test of significance

Difference is significant if:

(i) $\dfrac{M - X}{\sigma/\sqrt{n}} \geq 2$

(ii) $\dfrac{M - X}{S/\sqrt{n-1}} \geq t$ at 5% level

(iii) $\dfrac{X_1 - X_2}{\sigma/\sqrt{n}} \geq 2.8$

(iv) $\dfrac{X_1 - X_2}{S/\sqrt{n}} \geq t$ at 5% level

(v) $\dfrac{X_1 - X_2}{S/\sqrt{(1/N_1) + (1/N_2)}} \geq t$ at 5% level

Where,

X = sample mean, X_1 & X_2 = Individual sample means

n, N_1 and N_2 = no. of observation,

σ = population standard deviation,

M = population mean,

S = pooled standard deviation,

Application of normal distribution

Ex. 1: In a mill, count testing was performed and it was found 29.2 Ne against 30 Ne. 25 leas were tested. To know whether this difference is significant (real) or insignificant (not real)?

Ans: In this case, two things may arise i.e. (i) Population variance is known and (ii) Population variance is unknown.

When population variance is known, then apply the formula $\dfrac{M - X}{\sigma / \sqrt{n}} \geq 2$. If this relation is true, then it is significant, otherwise not.

In 2nd case, apply $\dfrac{M - X}{S / \sqrt{n - 1}} \geq t$ at 5% level.

Where,

X = sample mean,

n = no. of observations,

σ = population standard deviation,

M = population mean,

S = pooled standard deviation.

In the first case, M = 30s Ne

CV% = 3.0, then σ = 0.03 × 30 = 0.9, X = 29.2s Ne

Apply, $\dfrac{M - X}{\sigma / \sqrt{n}}$ i.e. $\dfrac{30.00 - 29.2}{0.9 / \sqrt{25}}$ = 4.44 which is greater than 2.0. Hence, the difference is significant.

Population variance means that CV% of 30's count as checked on day to day basis shows 3.0%.

In 2nd case σ is not known. Then σ as found by testing 25 leas are to be taken. Suppose it is 1.05, i.e. S = 1.05.

Apply, $\dfrac{M - X}{S / \sqrt{n - 1}}$ i.e. $\dfrac{30 - 29.2}{1.05 / \sqrt{25 - 1}}$ = 3.73.

Now, from 't' table, find out value of 't' at N=25 at 5% level. The value is 2.06.

As 3.73 is greater than 2.06. The difference of count is significant.

Ex. 2: A mill's nominal count is 100s Ne. After testing 25 numbers of leas it is found that the average count is 95s Ne. Department count CV% is 2.0. Is this count variation significant?

Ans: Here, apply $\dfrac{M-X}{\sigma/\sqrt{n}} \geq 2$ as population variance is known.

$M = 100s$, $X = 95$, $\sigma = 0.02 \times 100 = 2.0$.

So, $\dfrac{100-95}{2\sqrt{25}} = 12.5$ which is greater than 2.0. Hence, count variation is significant.

Ex. 3: From a group of 6 ring frames working on 100s Ne, two bobbins were taken from each ring frame for count testing. The average of 12 leas was 97.6s Ne. CV% of count is 3.0. Would you say that the wrapping count is actually or really coming coarser?

Ans: Here, apply $\dfrac{M-X}{\sigma/\sqrt{n}} \geq 2$

$M = 100^s$ Ne, $X = 97.6s$ Ne, $\sigma = 100 \times 0.03 = 3.0$, $n = 12$.

Hence, $\dfrac{100-97.6}{3/\sqrt{12}} = 2.77$ which is greater than 2.0. Hence, the difference is real or significant.

In the above example, if population CV% was not given, but count CV% of 12 leas came as 5.20. Would you say that coarser count was coming?

Here, apply $\dfrac{M-X}{S/\sqrt{n-1}} \geq t$ at 5% level

i.e. $\dfrac{100-97.6}{5.2/\sqrt{12-1}} = 1.53$. The value of '$t$' at $N = 12$ is 2.20 which is greater than 1.53. Hence, the difference is insignificant.

If instead of 5.20 count CV% had it been 3.0%, then what would be the conclusion?

$\dfrac{100-97.6}{3/\sqrt{12-1}} = 2.77$, which is greater than 2.20. Hence, it is significant.

Ex. 4: Suppose a draft wheel was changed in a ring frame and 25 leas were tested. Before changing the wheel, the count was 30.2^s Ne and after changing the wheel, it became 31.4^s Ne. Verify whether this change is significant or not? The department count CV% is 3.0.

Ans: Here, two cases may arise e.g.

(i) Population variance is known.

(ii) Population variance is unknown.

For, the first case, apply

$$\frac{X_1 - X_2}{\sigma / \sqrt{n}} \geq 2.8$$

If this relation is true, then it is significant, otherwise not.

Where, X_1 and X_2 = two mean,

σ = standard deviation of the population,

n = no. of tests,

For case (ii) apply

$$\frac{X_1 - X_2}{S / \sqrt{(1/N_1) + (1/N_2)}} \geq t \text{ at 5\% level.}$$

Now for case (i) $\dfrac{30.2 - 31.4}{0.906 / \sqrt{25}} = 6.55$ which is greater than 2.8. Hence, the change is significant. In this case, count CV% = 3.0 and population variance = 0.03 × 30.2 = 0.906.

Ex. 5: Two groups of ring frames working on 18^s Ne showed that the average count of two groups were 17.8^s and 18.3^s. 20 leas were tested for each group. Count CV% in both the cases was 4.0. To know whether both the groups are giving same or differing count?

Ans: Apply

$$\frac{X_1 - X_2}{\sigma / \sqrt{n}} \geq 2.8$$

Here, X_1 = 18.3, X_2 =17.8, σ = 18 × 0.04% = 0.72, N=20.

Hence, $\dfrac{18.3 - 17.8}{0.72 / \sqrt{20}} = 3.1$ which is greater than 2.8. Hence, two ring frame groups are differing from average count of 18^s.

Ex. 6: The $U\%$ of two samples of 40s combed yarn spun form the same mixing were found to be 16.5 and 15.9. From 16 ring frame bobbin one test/bobbin was done and the standard deviation of the two was found as 0.96 and 0.85. Are these two samples differing in $U\%$?

Ans: Apply

$$\frac{X_1 - X_2}{S / \sqrt{(1/N_1) + (1/N_2)}} \geq t \text{ at 5\% level.}$$

When pull variance,

$$S_2 = \frac{N_1 S_1^2 + N_2 S_2^2}{N_1 + N_2 - 2}$$

Hence, $X_1 = 16.5$, $X_2 = 15.9$, $N_1 = N_2 = 16$, $S_1 = 0.96$, $S_2 = 0.85$.
$S^2 = \{16 \times (0.96)^2 + 16 \times (0.85)^2\}/(16+16 - 2) = 0.87$.
$S = 0.93$

Hence, $\dfrac{16.5 - 15.9}{0.93 / \sqrt{\dfrac{1}{16} + \dfrac{1}{16}}} = 0.22$. The value of '$t$' at $n=16$ at 5% level

is 2.12.

Conclusion: 0.22 is less than 2.12. Hence, the difference is insignificant.

5.2   Common statistical tools used for discontinuous variables

Difference is significant if:

(i) $\dfrac{(\text{observed} - \text{expected})^2}{\text{expected}} \geq 4$

(ii) $\dfrac{(A - B)^2}{A + B} \geq 4$

Application of Poisson's and binomial distribution

Ex.7: End breaks in a ring frame was observed to be 10 breaks/spindle/h. Twist wheel was changed to reduce tpi by 3% and end breakages were found to be 15 breaks/spindle/h. Breakages were taken over a total period of 5 h in both the cases. Whether this increase in breakages was significant?

Ans: Before twist wheel change, total break = $10 \times 5 = 50$ (A).
After twist wheel change, total break = $15 \times 5 = 75$ (B).

Apply, $\dfrac{(A - B)^2}{A + B} \geq 4$

i.e. $\dfrac{(50 - 75)^2}{50 + 75} = 5$, which is greater than 4.0. Hence, it is significant.

Ex. 8: Neps in the card web was found to be 12.0 neps/100 sq. inch as an average based on 5 readings, but expected to be 10 neps/100 sq. inch as on average 5 reading. Is this difference is significant?

Ans: Apply, $\dfrac{(\text{observed} - \text{expected})^2}{\text{expected}} \geq 4$

i.e. $\dfrac{(12 \times 5 - 10 \times 5)^2}{10.5} = 2.0$, which is less than 4.0. Hence, this difference is insignificant.

Ex. 9: A mill maintain an average end breakage rate of 30 per 1000 spindle hour for 60s Ne. Due to change in mixing, the breakages rate was increased to the level of 40 per 1000 spindle hour. Has the change of mixing increased the breakage rate?

Ans: Here, apply

$$X^2 = \dfrac{(O - E)^2}{E}$$

i.e. $\dfrac{(40 - 30)^2}{30} = 3.3.$

The value of X^2 (Chi-square for 1 degree of freedom) is 3.84 which is greater than 3.3.

Hence, the change of mixing did not have any adverse impact on end breakages.

Ex. 10: In judging hairiness of two yarn samples A and B by winding them on black boards, the score obtained for A was 36 less than B out of 60 observations. Is yarn 'A' less hairy than B?

Ans: Here, apply

$\dfrac{(A - B)^2}{A + B} \geq 4$

$A = 36, B = (60{-}36) = 24$

i.e. $\dfrac{(36 - 24)^2}{36 + 24} = 2.4$

2.4 is less than 4.0. Hence, yarn A is not less hairy than B.

Ex. 11: A mill obtained 22 neps/100 sq. inches on 12 boards observed. The norm was 15 neps/100 sq. inches. Is this difference significant?

Ans: Apply

$$X^2 = \dfrac{(O - E)^2}{E}$$

Where, O = observed, E = expected.

Hence, $O = 22 \times 12$ boards $= 264$

$E = 15 \times 12 = 180$.

So, $\dfrac{(264 - 180)^2}{180} = 39.2$, which is greater than 4.0. Hence, the difference is significant.

Ex. 12: On changing the ring traveller form 3/o to 1/o, the end breakages of 30s Ne was reduced from 18.0/100 spl/h to 16/100 spl/h. The spindle hours observed for each trial was 2000 (20 h × 100 spls). Should we change the ring traveller to 1/o on all ring frame?

Ans: Apply, $\dfrac{(A - B)^2}{A + B} \geq 4$

Where, $A = 18 \times 20 = 360$, $B = 16 \times 20 = 320$.

So, $\dfrac{(360 - 320)^2}{360 + 320} = 2.35$, which is less than 4.0.

Hence, we should not change ring traveller to 1/o on all ring frames.

Significance test when the variances of two samples are known

$$\frac{(S_1)^2}{(S_2)^2} = \text{F at 5\% level}$$

Where, $(S_1)^2$ = Variance of the large sample,

$(S_2)_2$ = Variance of the smaller sample,

Ex. 13: A mill is producing 40^s count. On testing two samples drawn from two different spindles for tpi checking and calculating standard deviation, it was found that the standard deviation of 1st spindle was 1.30 based on 50 reading and that of 2nd spindle as 2.85 based on 60 readings. Are those two spindles varying in tpi?

Ans: Apply

$$\frac{(S_1)^2}{(S_2)^2} = \text{F at 5\% level}$$

$$= (2.85)^2/(1.31)^2 = 4.7.$$

Form F table, for $V_1 = 59$ and $V_2 = 49$, the value of F is 1.40 at 5% level which is less than 4.7. Hence, the difference is real or significant.

Ex. 14: After incorporating auto leveller (long term) in the high production card in a mill, the CV% of the card sliver decreased from 4.0 to 3.5. The mean hank of the sliver is 0.20. 40 readings were taken to measure CV% in both the cases. Is this difference statistically significant?

Ans: As two CV values are to be compared, apply F test.

$SD_1 = CV_1 \times$ mean/100 = $4 \times 0.2/100 = 0.008$ (Before attachment).

$SD_2 = CV_2 \times$ mean/100 = $3.5 \times 0.2/100 = 0.007$ (After attachment).

Now, $F = SD_1^2/SD_2^2 = (0.008)2/(0.007)2 = 1.31$.

The value of F at $n_1 = n_2 = (40 - 1) = 39$ at 5% level is 1.50 which is greater than 1.31. Hence, there is no improvement in CV%.

5.3   To find the population mean from a small sample

Ex. 16: Eleven ring bobbins were tested for lea strength. The mean is 50 lbs and the standard deviation is 6 lbs. Estimate 95% confidence interval for the mean lea strength of the population.

Ans: Standard error (S.E.) of the mean $= \dfrac{S.D}{\sqrt{n}}$

Here S.E. $= \dfrac{6}{\sqrt{11}} = 1.8$.

For 95% confidence interval, the confidence limit is std. mean $\pm (t \times S.E)$

Where, t is taken at 5% level.

In the above case, the value of t for $n = 11-1$ or 10 is 2.228.

So, confidence limit = 50 lbs $\pm (2.228 \times 1.8)$

= 50 lbs $\pm$ 4 lbs.

Hence, based on 11 reading for lea strength, the lea strength will vary from 46 lbs to 54 lbs at 95% confidence level.

5.4   Accepted tolerance limit for deviation

It is given by, avg. value $\pm 3\sigma/\sqrt{n}$

Where, σ = standard deviation,

                     n = sample size.

Application: Suppose avg. count is 29.8 and CV% = 1.80 based on 50 readings.

The deviation which is to be tolerated is $29.8 \pm \dfrac{3\sigma}{\sqrt{n}}$ i.e. $29.8 \pm \dfrac{3 \times 0.53}{\sqrt{50}}$

ie. 29.8 ± 0.22

i.e. Tolerated count will be 29.58s to 30.02s

5.5   Error in estimation of CV%

It is given by CV% $\pm \dfrac{2CV\%}{\sqrt{n}}$

Application: CV% = 1.80, $n = 50$.

Error in estimation of CV% will be $1.8 \pm \dfrac{2 \times 1.8}{\sqrt{50}} = 1.8 \pm 0.50$ i.e. CV% will vary from 1.30 to 2.30.

5.6   Number of reading to be taken with a given CV% and 1% error at 95% confidence limit

It is given by

$$n = 4 \times (CV)^2/(Error)^2$$

Application: suppose, count CV% = 2.0 then how many reading are to be taken with 1% error?

It is given by $4 \times (2)2/(1)2 = 16.0$.

Practical application of above formula:

Suppose CV% of yarn count of any ring frame was found to be 3.0 based on 12 readings. Now, how many reading we should take for 1% error?

It is given by: $4 \times (3)^2/(1)^2 = 36$.

So, we should go for 36 numbers of tests to get 1% error in our test result for the above case.

5.7   To find out the number of breaks study (spl. hour) reqd. to be done in ringframe to arrive at a predetermined error

Application:

How many spindle hours study is required with an error of 5% with an average breakage in a ring frame as 4/100 spindles hour?

Ans: We are to apply the formula, $E\% = \dfrac{200}{\sqrt{n}}$ ...(1)

Where, E is the error% and n is total number of breaks.

From Eq. (1), $n = 40,000/(E\%)^2$

Hence, for our above problem, total number of breaks to be observed = $40000/(5)^2 = 1600$.

Breakage rate is 4/100 spindles hour.

For 5% error, spindle hours observation will be $= \dfrac{1600}{4} \times 100 = 40{,}000$

i.e. for 100 spindles, 400 h study is required, and for 1008 spindles in a ring frame, 40 h of study is required.

5.8   Number of snap studies required to find out the predetermined efficiency of a group of machines

We have to apply the formula, $n = 4p\,(1-p)/(E\%)^2$

Where,

p = predetermined efficiency,

E = error,

N = number of studies.

Application:

Ring frame efficiency in a spinning mill containing 50 ring frames is to be estimated through snap study with an error (%) not more than 1. Preliminary study indicates the average efficiency could be around 95%. What should be the number of snap studies required in this case?

Ans: Here, $p = 0.95$, $1 - p = 0.05$, E (%) = 1.0.

$N = 4 \times 0.95 \times 0.05/(.01)^2 = 1900$.

Therefore no. of snap studies reqd. $= 1900/50 = 38$.

5.9   Confidence limits for number of events (Poisson's)

During observation on a ring frame for 4 h, the total numbers of ends down were found 49. Set the confidence limit of ends down of 4 h.

Ans: Here we have to apply the formula for the confidence limit as $r \pm 1.96\sqrt{r}$

Where, r = number of ends down.

So, confidence limits $= 49 \pm 1.96\sqrt{49} = 49 \pm 14$.

i.e. During 4 h of study, ends down will vary from 35 to 63 (49 ± 14).

Yarn and package defects

Yarn defects arise from raw materials (wrong selection or poor quality), faults in machinery settings, improper maintenance, etc. There are also different faults in package formation due to defective winding. Both yarn and package defects lead to different types of fabric faults (woven and knitted) which are objectionable.

6.1 Yarn defects

1. Slubs

A soft thick place or lump in yarn showing less twist at that place.

Effect:
- More end breaks in the next process.
- Damaged fabric appearance.
- Shade variation in dyed fabrics.

Causes:
- Poor carding.
- Defective ring frame drafting. As the spinning draft is increased or lowered from the optimum draft the frequency of slubs may increase [1].
- Higher break draft could also be a reason for higher slub formation during drafting at ring frame [2].
- Lack of adequate pressure in back zone.
- Excessive spindle speed may causes increase in the number slub [3].
- Improper apron spacing may also cause higher slub formation [3].
- More short fibres in mixing.

Remedies:
- Better fibre individualisation at cards to be achieved.
- Use of proper total and break draft.
- Optimum top roller pressure in back zone.
- Setting at ring frame to be maintained.

2. Neps

- Yarn containing rolled fibre mass, which can be clearly seen on black board at close distance; measurable on Uster imperfection Indicator.

Effect:

- Damaged fabric appearance, particularly in woven fabrics. In knitted fabric, its presence is submerged under loop to a large extent.
- White specks in dyed fabrics, as neps are formed mostly due to immature fibres whose dye pick up is less.

Causes:

- Poor carding.
- Too wide a flat setting produces significantly higher level of neps than the normal setting, due to loss of control over the fibres [4–6].
- Immature cotton in mixing.

Remedies:

- Better fibre individualisation at cards to be achieved.
- Optimum closer setting between flat and cylinder.
- Front stationary flats setting to be made closer.
- Flat waste to be increased.
- Card hank to be made finer.
- Metallic wire condition should be proper.

3. Snarl

The snarl is due to yarn torque and takes place during the torsional buckling effect where the yarn retires into itself and simultaneously gets twisted in the opposite twist direction.

Effect:

- In weaving, snarls in weft yarns appear when the shuttle returns to the shed from the box. Some of these snarls are entrapped into the cloth and do not open out even when weft is subsequently tensioned as picking proceeds making necessary the cloth mending for their removal [7].
- As the hanks made from highly twisted yarns are removed from the reel swift, they shrink and form numerous snarls, causing great trouble during their positioning on the rods of the hank dyeing machine [8].
- Moreover, during the winding of these dyed yarn hanks to cones, the formed snarls can cause snatching, tension peaks and yarn breakage.

- Damaged woven fabric appearance as snarl appears as short thick place in the fabric.
- If a twist-lively yarn is used for knitting, the resultant loop will no longer be symmetric because of the varying induced torsional strain in the yarn [9].

Causes:

- Higher than normal twist in the yarn.
- Presence of too many long thin places in the yarn.
- Post-spinning operations are mainly responsible to produce snarls; sources are:
    i.   Manual cone winding section – Snarls are produced at manual knotting portion.
    ii. Cheese-winding section – After knotting, both ends are to be released with proper tension by pulling the ends otherwise snarls will be formed at knotted portion.
    iii. Ring-Doubling – Snarls will be formed if two ends of creeled cheese are split apart causing differential tension in the yarn.
    iv. T.F.O – Improper functioning of brake capsules will result in snarl formation.

Remedies:

- Optimum twist to be used for the type of cotton processed.
- Drafting parameters to be controlled minimise thin places in the yarn.
- The yarn to be conditioned.
- Correct tension weights and slub catcher settings to be employed at winding.
- Use of steel balls in place of capsules in TFO with setting of reserve will reduce incidence of snarl formation.

4. Thick and thin places

Measurable by Uster Imperfection Indicator and observable on appearance.

Effect:

- More ends down in downstream processing.
- Can cause weft bars, diamond barring effects [10], moiré effects, weft stripes or rings in the resulting fabric [11].
- Disturbing appearance in the fabric.

Causes:

- Eccentric top and bottom rollers.
- Insufficient pressure on top rollers.
- Worn and old aprons and improper apron spacing.
- Improper meshing of gear wheels.
- Mixing of cottons varying widely in fibre lengths and use of immature cottons.
- Improper total and break draft on ring frame.
- Too high shore hardness of top cots.
- Too long buffing intervals of top cots.
- Creel stretch.

Remedies:

- Eccentric top and bottom rollers to be avoided.
- Top arm pressure checking schedules to be maintained strictly.
- Wide variation in the properties of cottons used in the mixing to be avoided.
- Better fibre individualisation at cards to be achieved.
- Correct spacers to be used.
- Shore hardness of 65°and75° for cotton and synthetic to be used to provide nip area rather than nip point control resulting better grip on fibres.
- Buffing should be done on regular intervals. Buffing creates micro holes on top cots giving better grip between top cots and bottom fluted rollers.

5. Soft yarn

Yarn which is weak indicating lesser twist.

Effect:

- More end breaks in subsequent processes.
- Shade variation in dyed fabrics.

Causes:

- Slack tapes dirty jockey pulleys.
- Improper bobbin feed on the spindles.
- Less twist in the yarn due to missing of spindle button.

Remedies:

- Vibration of bobbins on the spindles due to oversize inner diameter of empty bobbin to be avoided.

- Periodic replacement of worn rings and travellers to be effected.
- Spindle button missing to be taken care of.
- Loose tape and jockey pulley spring to be attended.

6. Oil stained yarn

Yarn stained with oil.

Effect:

- Damaged fabric appearance.
- Occurrence of black spot in fabric.

Causes:

- Careless oiling in the moving parts.
- Piecings made with oily or dirty fingers.
- Careless material handlings.

Remedies:

- Appropriate material handling procedures to be followed.
- Oilers to be trained in proper method of lubrication.
- Clean containers to be utilised for material transportation.

7. Crackers

Crackers are short coil like places in the yarn which are characterised by the cracking sound produced when the yarn is straightened by pulling due to sudden rupture of fibres curled around the yarn. Cracker formation is due to the presences of few long fibres in the roving, longer than nipping distance between front and middle rollers, which are stretched and extended during drafting as both ends are gripped by both front and middle rollers. These long fibres get contracted after being released and take adjacent fibres along with them which are twisted forming crackers.

Effect:

- More breaks in winding.
- More noticeable in polyester and cotton blended yarns due to high extensibility of polyester fibres.

Causes:

- Mixing of cottons of widely differing staple length.
- Closer roller settings.
- Eccentric top and bottom rollers.

Remedies:

- TPI of yarn to be made lower till the CSP does not fall.
- Heavier ring traveller to be used.
- Spindle speed to be increased.
- Break draft to be increased.
- Ring rail speed to be increased.
- Roving tpi to be lowered.

8. Bad piecing

Unduly thick piecing in yarn caused by over end piecing.

Effect:

- More end breaks in subsequent process.
- Increase in hard waste.

Causes:

- Wrong method of piecing and over end piecing.
- Lack of skill of ring frame tenter.

Remedies:

- Tenters to be trained in proper methods of piecing.
- Excessive end breaks in spinning to be avoided.

9. Oily slub

Slub in the yarn stained with oil.

Effect:

- More end breaks in the ensuring process.
- Damaged fabric appearance.
- Shade variation in dyed fabrics.

Causes:

- Accumulation of oily fluff on machinery parts.
- Poor methods of lubrication in preparatory processes.
- Negligence in segregating the oily waste from process waste.

Remedies:

- Yarn contact surfaces to be kept clean.
- Oilers to be trained in correct procedures of lubrication.
- Proper segregation of oily waste from process waste.

10. Kitty yarn

Presence of black specks of broken seeds, leaf bits and trash in yarn.

Effect:
- Damaged fabric appearance.
- Production of specks during dyeing.
- Needle breaks during knitting.

Causes:
- Ineffective cleaning in blow room and cards.
- Use of cottons with high trash and too many seed coat fragments.

Remedies:
- Cleaning efficiency of blow room and cards to be improved.

11. Hairiness

Protrusion of fibre ends from the main yarn structure.

Effect:
- Uneven fabric surface.
- High standard deviation of hairiness produce barriness in knitted fabrics.
- Beads formation in the fabric in the case of polyester/cotton blends.

Causes:
- Use of cottons differing widely in the properties mainly in micronaire or denier value in the same mixing.
- Use of worn rings and lighter travellers.
- Maintaining low relative humidity, closer roller settings and very high spindle speeds.
- Grooved lappet hook in ring frame.
- Rotating yarn touches separator due to high ballooning in ring frame.

Remedies:
- Use of travellers of correct size and shape and rings in good condition to be ensured.
- Periodic replacement of travellers.
- Roller settings to be maintained.
- Optimum relative humidity to be maintained in the spinning room.
- Wide variation in the properties of cottons used in the mixing to be avoided.

12. Foreign matters

Metallic parts, jute flannel and other similar foreign matters spun along with yarn, polyethylene filament, jute twines, plastic bags.

Effect:

- Breaks during winding.
- Formation of holes and stains in cloth.
- Damaged fabric appearance.

Causes:

- Improper preparation of mixings.
- Improper bale sorting or picking by manual pickers.

Remedies:

- Removal of foreign matters (such as jute fibres, colour cloth bits) to be ensured during preparation of mixing.
- Installation of permanent magnets at proper places in blow room lines to be ensured.

13. Spun in fly

Fly or fluff either spun along with the yarn or loosely embedded on the yarn.

Effect:

- More breaks in winding.

Causes:

- Accumulation of fluff over machine parts.
- Fanning by workers.
- Failure of overhead cleaners.
- Malfunctioning of humidification plant.

Remedies:

- Machinery surfaces to be kept clean by using roller pickers.
- Fanning by workers to be avoided.
- Performance of overhead cleaners and humidification plants to be closely monitored.

14. Cork screw yarn

A spiral or corkscrew yarn is a plied yarn that displays a characteristic smooth spiralling of one component around the other. When two yarns of widely varying counts are plied together, the coarser yarn remains straight and normal yarn coils around the coarser yarn thus forming the corkscrew effect.

Effect:
- Breaks during winding.

Causes:
- Feeding of two ends (instead of one) in ring frame.
- Lashing-in ends in ring frame/simplex.

Remedies:
- Tenters are to be trained in piecing methods(or) practices.
- Pneumafil ducts to be kept clean and properly set.
- Separators in simplex to be used preferably serrated holes.

6.2   Package defects

1. Slough off

Coils of yarn coming out of the ring cops in bunches at the time of unwinding

Effect:
- Increase in end breaks.
- Higher yarn waste.

Causes:
- Improper ring rail movement.
- Worn builder cam.
- Loose package and excessive coils in the package.
- Soft build of cops.
- Improper empties fit on the spindles and slack tapes.

Remedies:
- Ring rail movement to be set right.
- Optimum ratio of winding: bindings coils and optimum chase length to be maintained.

2. Ring cuts

Damaged layers on the surface of the ring cops.

Effect:
- Excessive breaks during winding.
- More hard waste at winding.

Causes:
- Spindle or empty cops wobbling.
- Use of lighter travellers and incorrect ratchet wheel.
- Movement of spindles to the rings not concentric.

Remedies:
- Worn spindles to be replaced.
- Improper fit of empty cops with spindles to be avoided.
- Gauging of spindles with rings to be properly carried out.
- Use of right type traveller and ratchet wheel to be ensured.

3. Low cop content

Yarn content in the cop is less.

Effect:
- Efficiency loss in ring frame.
- Drop in winding efficiency.
- More knots for a given length of wound yarn.

Causes:
- Underutilisation of bobbin height.
- Lower number of coils/inch.
- Higher chase length.
- Cop bottom bracket properly not set.
- Improper selection of ratchet.
- Ratchet pawl pushing number of teeth/movement in the ratchet wheel.
- Spinning empties wall thickness is high.

Remedies:
- Optimum chase length, coil spacing and wall thickness of empty cops to be ensured.
- Ratchet/pawl movement to be properly set.
- Free space of only 7.5 mm to be maintained at the top and bottom of the cop.
- Free space only 0.75 mm only to be maintained between full cops and the ring.

4. Improper bobbin build

Step-like appearance of the cop.

Effect:
- Slough-off during doffing/winding.
- More breaks during unwinding (due to slough off).
- Higher hard waste in winding.

Causes:

- Improper combination of ratchet and pawl.
- Jerky ring rail movement (pocker rod movement to check).

Remedies:

- Ratchet and ratchet/pawl movement to be accurately arrived at taking into consideration.
- Count of yarn, ring diameter and chase length.
- Lubrication of pocker rods at appropriate intervals to be carried out.

5. Stitching on cone

Ends not laid properly on the cone at reversal of yarn path.

Effect:

- More end breaks in the subsequent process.
- Excessive yarn waste.

Causes:

- Vibrating and wrongly set cone holder.
- Yarn coils wrapped round the base of cone holder.
- Traverse restrictors fixed at incorrect position.
- Improper alignment of tension brackets with the drum.

Remedies:

- Maintenance cone winders to be good.
- Cone holder settings and alignment of tension.
- Brackets with drum to be carried out as frequently as possible.

6. Ribbon wound cone

Formation of ribbon like structure on the circumference of the cone.

Effect:

- High level of slough off during unwinding.
- Excessive yarn waste.
- Uneven dye pick up in the case of dye packages.

Causes:

- Winding spindle not revolving freely.
- Cone holders incorrectly set.
- Defective settings of cam switch.
- Lint accumulation in builder cam groove.

Remedies:
- Over hauling of cone winders to be periodically carried out.
- Anti-ribboning mechanism to be checked at frequent interval.
- Free movement of the cone holders to be ensured by proper lubrication.

7. Soft build cones

Unduly soft structure of cone.

Effect:
- Overall density of package is lower.
- Soft packing either at the base or at the nose of cones.

Causes:
- Improper alignment of winding spindle to the winding drum.
- Insufficient unwinding tension.
- Inadequate cradle loading.

Remedies:
- Unwinding tension to be maintained at 6–8% of single yarn strength.
- Cradle pressure to be maintained to the optimum level.

8. Bell shaped cone

Cones which are tightly built at centre, presents a shape of bell.

Effect:
- Excessive breaks during subsequent processes.

Causes:
- High yarn tension during winding.
- Cone holders incorrectly set to the winding drum.
- Damages in paper cone centre.

Remedies:
- Quality of cones to be checked at that time of procurement.
- Optimum unwinding tension to be maintained.

9. Nose bulging

Bulging of bunches of the yarn at the nose of the cones.

Effect:
- Slough during warping/unwinding.
- Excessive yarn waste in next process.

Causes:
- Improper setting of cone holders to the winding drum.
- Damaged nose of the paper cones.

Remedies:
- Periodical inspection of settings in winding machines.
- Tenters to be instructed to adopt correct work practices.
- Avoiding usage of damaged paper cones.

10. Collapsed cone

In this type of defect, the structure of the package itself gets collapsed.

Effect:
- Excessive breaks during warping.
- Tend to generate a high level of hard waste.

Causes:
- Use of poor quality/damaged cones.
- Poor system of material handling.
- Maintaining non optimum unwinding tension.

Remedies:
- Using of poor quality/damaged paper cones should be avoided.
- Winding tenters should be trained by proper work methods.
- Proper material handling devices such as cone transport trolleys to be used.
- Cone inserts to be used for paper cones.

11. Ring shaped cone

Formation of ring shaped bulge across the cross section of the cone.

Effect:
- More end breaks in the subsequent processes.
- Slough off during unwinding.

Causes:
- Incorrect setting of the cone holder.
- Wrong placement tensioners in the tensioning assembly.
- Traverse of yarn affected due to defects in the grooves of the drum.

Remedies:

- Due to replacement of defective drums and stop motion wires to be ensured.
- Periodic inspection of cone holder settings and tension assembly to be carried out.

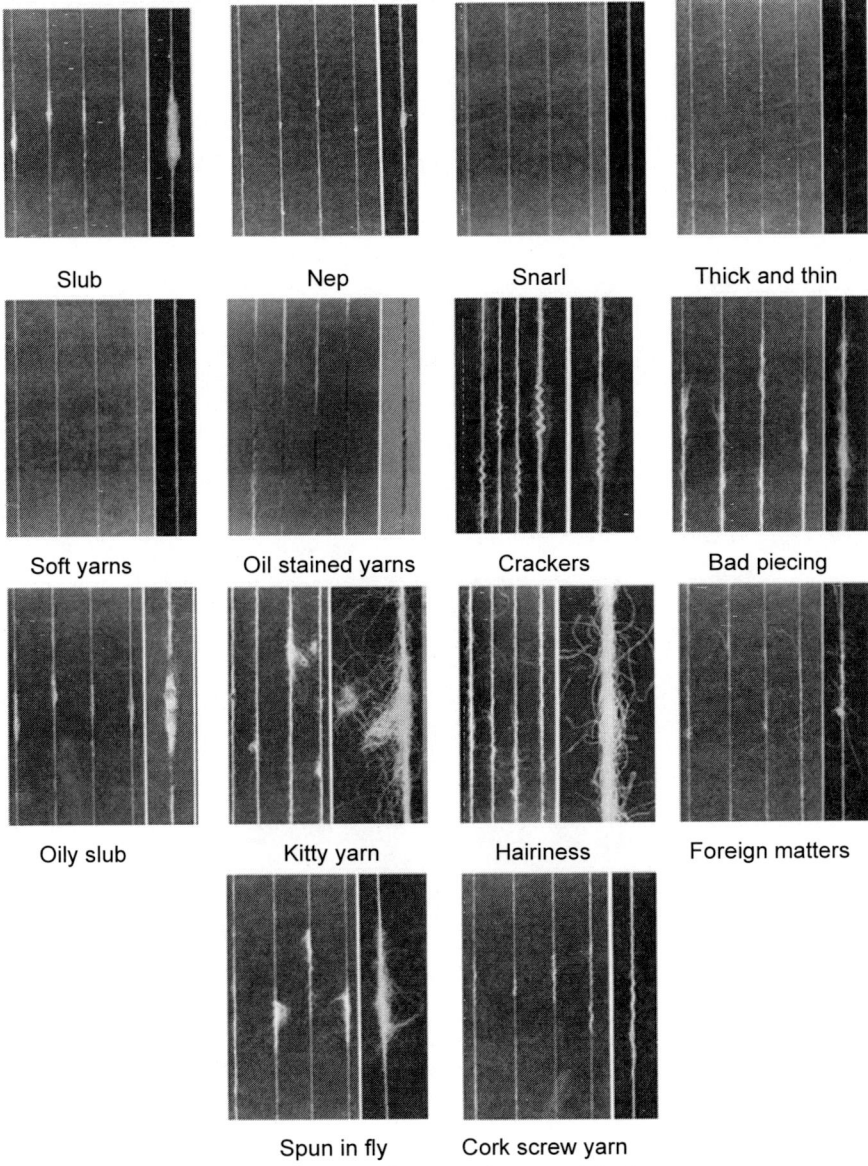

| Slub | Nep | Snarl | Thick and thin |

| Soft yarns | Oil stained yarns | Crackers | Bad piecing |

| Oily slub | Kitty yarn | Hairiness | Foreign matters |

| Spun in fly | Cork screw yarn |

Figure 6.1: Different major yarn defects

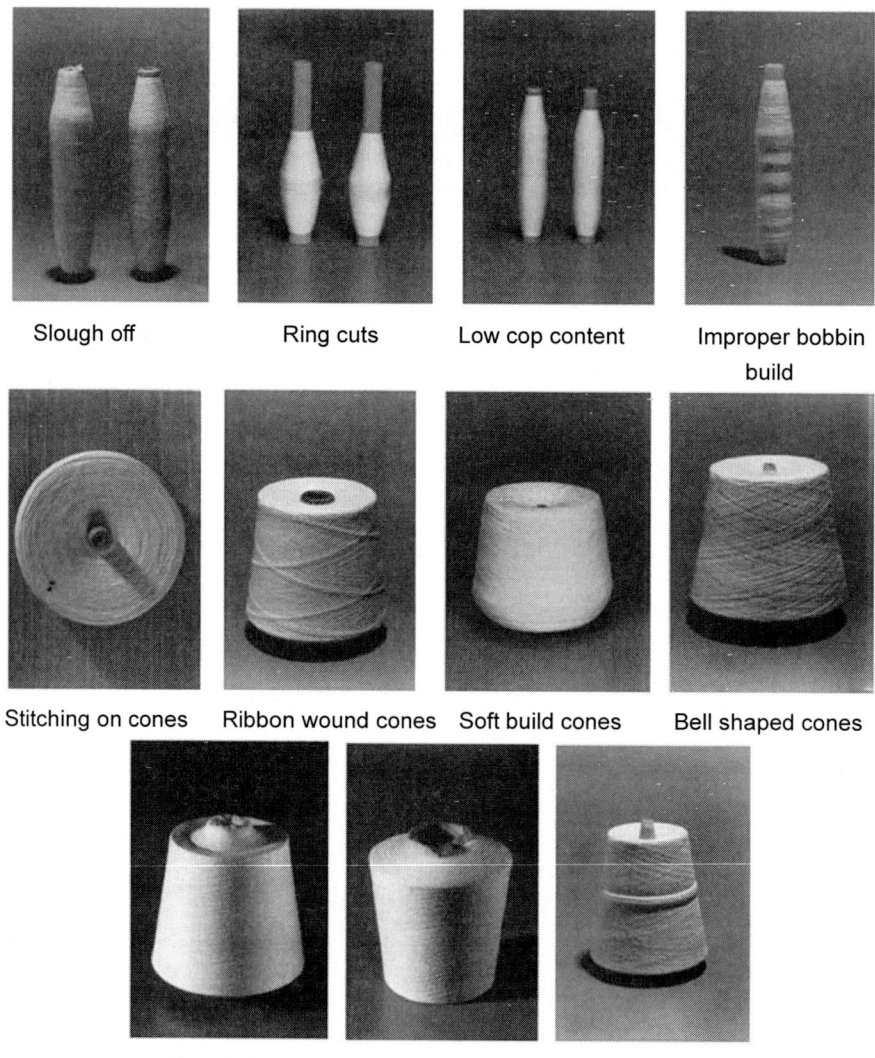

Figure 6.2: Different major package defects

6.3   References

1. Balasubramanian, N., 1975, The Effect of Top-Roller Weighting, Apron Spacing, and Top-Roller Setting Upon Yarn Quality, Text. Res. J.,45 (1975)322.

2. Grover, J. M. and Subramanian, T. A., Proceedings, 15th Technological Conference of ATITA, BTRA and SITRA, p. 10.

3. Pillay, K.P.R and Hariharan, R., Proceedings, 25th Technological Conference of ATIRA, BTRA and SITRA, p. 1.

4. Harrison, R.E. and Bargeron, J.D., Comparison of several nep determination methods, Text. Res. J., February, p. 77–79, 1986

5. Lord, P.R., Handbook of Yarn Production – Technology, Science and Economics, T.T. Institute, ed., Woodhead Publishing Limited, Cambridge, 2003, p. 492.

6. Pattabhiram, T.K., Essential Element of Practical Cotton Spinning,(4th edn.) Somayya Publication, 1997.

7. Beevers, H., Practical Spinning on the Bradford system, National Trade Press, London, 1954, 45.

8. Primentas, A.,Ind. J. Fibre Text. Res., 28, March 2003, p. 23–28.

9. Tao, X.M., Lo, W.K., Lau, Y.M., Torque balanced singles knitting yarns spun by unconventional systems, Part I: cotton rotor spun yarn, Text. Res. J., 67(10), 1997, pp. 739–746.

10. Catling, H., Some effects of sinusoidal periodic yarn thickness variations on the appearanceof woven cloth, J. Text. Inst., 49 (1958) pp. T232–T246.

11. Furter, R., Evenness testing in yarn production: Part I & II, The Textile Institute,Manchester, 1982, pp. 53–73 and15–39.

In the last 25 years global demand for fibres has grown by 124% to 88 million tonnes. Man-made fibres contributed 90% of the growth in fibre consumption in the last 25 years with volumes rising from 19 million tonnes in 1990 to 63 million tonnes in 2015. Consumption of cotton increased from 19 million tons to 24 million tons in this period.

Since the adoption of many man-made fibres by the textile industry, the blending of man-made fibres with each other or with natural fibres has become popular. Blending may be defined as the mixing of two or more types of fibres so that the resulting mixture has characteristics of the average of the component items [1]. Speakman [2] has given three main reasons for blending. The first is to produce a cheaper product. It should, however be emphasised that low cost is not necessarily a desirable end; cost in relation to performance is the basic criterion of merit. The second reason is the correction of defects. Many fibres have some inherent undesirable characteristics which can be minimised by blending with some other fibres. The third and, perhaps, the most important reason for blending is the creation of new and more desirable effects. Blending different types of fibres is a widely practiced means of enhancing the performance and the aesthetic qualities of a fabric. Blended yarns from natural and man-made fibres have the particular advantage of successfully combining the good properties of both fibre components, such as comfort of wear with easy care properties. These advantages also permit an increased variety of products to be made, and yield a stronger marketing advantage [3, 4]. Polyester fibre which is one of the most preferred among synthetic fibres is widely used both alone and blend with other fibres [5]. One of these fibres is cotton. Cotton, as a natural cellulosic fibre, has a lot of characteristics, such as comfortable soft hand, good absorbency, colour retention, prints well, machine-washable, dry-cleanable, good strength, drapes well, easy to handle and sew [6].

7.1   Type of man-made fibres used

In man-made fibre spinning there are mainly 3-types of fibres which are used individually or in combination of one another to produce synthetic/blended yarns. These fibres are:

(i) Polyester,

(ii) Viscose,

(iii) Acrylic.

7.1.1   Fibre specification

These fibres are manufactured in different deniers and cut lengths and spun into yarn either with 100% any one type of fibre or in combination (blending) of any two, like polyester/viscose, polyester/acrylic, and acrylic/ viscose. Amongst all above three types of fibres, the usage of polyester fibres is maximum due to its versatile properties and lower price. There is also a big application of polyester in home textile. Table 7.1 shows the denier, cut length, blending partner and end use of polyester fibres for apparel uses. For blending polyester with viscose fibres or cotton fibres, one should use T-10/T-15 (tenacity at 10% elongation or 15% elongation) for viscose fibres and T-7 (tenacity at 7% elongation) for cotton fibres for elongation balance and compatibility in matching between two components of fibres.

Table 7.1: Type of polyester fibres used (100% or blends) for different apparel fabrics

Denier	Cut length (mm)	Blended with	End user
1.4/2.0/2.5 TBL (all medium tenacity)	38, 44, 51	Viscose	Suiting, shirting
1.2/1.4 (High tenacity)	32, 34	Cotton	Shirting's
0.8/1.0	38, 44, 51	Viscose	Ladies blouse/sarees
1.2DSuper high tenacity	40, 44	100%	Sewing threads
6D(Hollow), 7D (Conjugate)	64	Coarser denier viscose	Suiting's with worsted effect, shawls, etc.

Table 7.2 gives a brief idea about denier and cut length of polyester fibres for application in home textiles (Reliance Products Recron).

Table 7.2: Denier and cut length of polyester fibres for different home textiles

Denier	Cut length (mm)	Products attributes	Application
0.8 and1.0 (Micro)	32, 40, 44, 51	Soft feel and hand-blends with cotton and viscose	Premium quality bed sheets, pillow cases, cushion cover, quilts, etc.
Cot look 1.2, 1.4	38, 40, 44, 51	Looks and feels like cotton	Bed sheets, pillow cases, cushion cover
Super dye 1.2, 1.4, 2.0	32, 38, 40, 44, 51	Bright and brilliant colours, soft and gentle fell, better pilling performance	Bright and vibrant curtain, table cover

Contd...

Contd...

Denier	Cut length (mm)	Products attributes	Application
Pre coloured 1.4	44, 51	High consistency of shade, high all round fasten	Bed sheets, pillow cases, cushion cover
Carpet fibre 15.0 (circular and trilobal)	190 (V.C.)	Exceptionally bright lustre, good bulk and uniform dyeing and TBL cross section for sparkle	Spun yarn for women and hand tufted nonwoven needle punched carpet
Low elongation fibre 1.4	32, 38, 40, 44	Low elongation in yarn	Carpet backing
Spun lace 1.2, 1.5	38, 51	Workable at high speed carding. Blends well with cellulosic fibre	Dry and wet wipes. Towels, medical gowns

Table 7.3 shows the application of viscose fibres as a blend component with other fibres for different inner and outer wears (Source: Technical bulletin of Birla viscose)

Table 7.3: Application of viscose fibres in blends (Source: Technical bulletin of Birla viscose)

Product	Blend %	Inner wear count (Ne)	Outer wear count (Ne)	Salient features
Viscose	100%	34s, 40s	20s, 30s	Soft, high absorbency, high lustre and drape
Cotton/viscose	55/45, 75/25	34s ,40s	20s, 24s, 30s	Cooler on skin
Acrylic/viscose	50/50, 65/35		20s, 24s, 30s	Light and bulky, high absorbency and fast shades
Polyester/viscose	65/35, 48/52		20s, 24s, 30s, 34s	Durable, soft, excellent drape
Viscose/nylon	50/50, 70/30, 83/17		24s, 30s	Lighter, comfort, good elasticity and drape
Wool/viscose	20/80		20s, 24s	Softer, adds to drape, bright colour, versatile to all weather

Range of denier and cut length of different types of viscose fibres are given in Table 7.4 (Source: Technical bulletin of Birla viscose)

Table 7.4: Viscose fibre range (Birla viscose)

Type	Denier	Cut length (mm)
Bright bleached	0.8–12.0	32–120
Dull/semi dull	1.2–12.0	32–120
Spun shades	1.5–12.0	32–120
Anti-bacterial	1.5	38

Contd...

Type	Denier	Cut length (mm)
Chlorine free	1.5–12.0	38–120
Grassi soft (hollow flat)	1.5–6.0	38–120
Grassi soft	0.8	38–44

Grassi soft has a wide application for blending with polyester, cotton and polyester/cotton in the count range of 20s, 24sand30s. These blended yarns give a very soft feeling, excellent drape and cool to the skin.

Dope dyed viscose fibres are also used for polyester/viscose blended yarn. Table 7.5 shows different colours available with corresponding shade nos. Deniers available are 1.5D/3D/4D. Cut length for 1.5D fibres are 38, 44 and51 mm. For coarser denier, cut lengths are 76, 64 and60 mm (Manufactured by Grasim, Nagda).

Table 7.5: Dope dyed viscose

Sl. no.	Shade no.	Colour
1.	1450	Black
2.	1760	Navy
3.	2746	Coffee
4.	1405	Maroon
5.	1455	Dark royal blue
6.	3571	Red
7.	2925	Medium bright blue
8.	1448	Green
9.	3516	Mustard
10.	1989	Bottle green
11.	1320	Bhagwa
12.	4141	Chocolate brown
13.	1700	Olive
14.	4029	Light brown
15.	3092	Slate
16.	3323	Light brown
17.	2747	Medium grey
18.	2104	Light khaki
19.	3003	Silver shadow
20.	3732	Beige
21.	2247	Light grey
22.	3777	Mist grey
23.	4365	Light mist grey
24.	3475	Sky blue
25.	3474	Deep beige
26.	3706	Cream
27.	3517	Light olive

7.1.2   Use of acrylic fibres

- 1.5D (Normal) for hosiery yarn: count range 24s–40s,
- 3D, 5D, 8D for bulking: Count range from 4s to 20s,
- 0.9D micro denier for special soft feeling: Count is mostly 40s.

Acrylic high bulk fibres are available in two categories:

(i) Dry spun.

(ii) Wet spun.

Fibres shrinkages are different in these two categories, Wet spun fibres having more shrinkages than dry spun fibres. For high bulk yarns, actual count and nominal count differs due to shrinkage of fibres.

$$\text{Actual count} = \frac{\text{Nominal metric count}}{169} \times \frac{100}{100 - \text{Shrinkage (\%)}}$$

Shrinkage (%) of both wet and dry spun acrylic fibres after bulking and dyeing are given in Table 7.6.

Table 7.6: Shrinkage (%) of acrylic fibres

Type	Shrinkage % after bulking and coning	Shrinkage % after dyeing and coning
Wet spun fibre	24–25	21.5–22.5
Dry spun fibre	14–15	12–13

Few dry and wet spun acrylic fibres used in the industry are:

Dry spun

Supercryl – HSSF(L1), HSSF (L2) – IPCL

Indalon TS (Indian Acrylic)

Cashmilon FKBR (L2)

Finnel – F101 (H)

Wet spun

Indacryl, HLO (IPCL)

Pasupati HS

Vonnel –V

Thai Texlan

Cashmilon – FKBR(L1)

7.1.3   Fibre length and fibre length to denier ratio

Selection of denier and fibre length to denier ratio (Table 7.7) are two important factors deciding ease of processing and ultimate yarn properties, as discussed below:

Table 7.7: Fibre length and fibre length to denier ratio

Fibre length in cm (*L*)	Denier (*D*)	Ratio (*L/D*)	Remarks
3.8	1.2	3.17	Comfortable process
5.1	1.0	5.1	Difficult to process impacting yarn quality

$L/D > 3$ leads to:

(i)   More chances of beater, cylinder getting loading.

(ii)   Increase in nep generation and hence deterioration in yarn appearance.

$L/D < 3$ leads to:

(i)   Reduction in yarn strength.

(ii)   Increase in end breakages.

Finer fibres exhibit:

(i)   Lower bending/torsion rigidity.

(ii)   Better drape ability.

(iii)   Soft feel.

Coarser fibres exhibit:

(i)   Harsh and rough feel.

(ii)   More bending and torsion rigidity.

Following range of deniers are suitable for different types of fabrics:

- 1.0D to 1.5D suitable for shirtings/sarees.
- 1.5D to 3D Suitable for medium and heavy fabric.

7.1.4   Acrylic fibre mixing for manufacturing acrylic bulk yarn

1.   Metric count 2/28

Mixing:

3.3D × 60 mm regular (v-17): 50%.

3.3D × 60 mm H.B (v-17): 50%.

Shrinkage after dyeing and coning is 22%.

Ring frame count to be kept $(Ne) = \dfrac{28}{1.69 \times 0.78}$

i.e., $21.2 \pm 0.2 = 21.0s$ to $21.4s$

2.   Metric count 2/25 (Indacryl)

Mixing:

3D × 64 mm IPCL regular: 60%.

$3D \times 64$ mm IPCL H.B: 40%.

Shrinkage after dyeing and coning is 22%.

$$\text{Ring frame count } (Ne) = \frac{25}{1.69 \times 0.78} = 19.0s.$$

i.e. $19 \pm 0.2 = 18.8s$ to $19.2s$

3. Metric count 2/25(Sparkle)

Mixing:

$3D \times 64$ mm IPCL (L1): 40%.

$3D \times 64$ mm IPCL (Regular –non shrinkable): 42%.

$10D \times 64$ mm JKTBL polyester: 18%.

Shrinkage after dyeing and coning is 12%.

$$\text{Ring frame count } (Ne) = \frac{25}{1.69 \times 0.88} = 16.8s.$$

i.e. $16.8s \pm 0.2 = 16.6 - 17.0$.

4. Metric count 2/32 (Supercryl)

Mixing:

$3D \times 64$ mm (L2 Supercryl): 50%.

$3D \times 64$ mm IPCL (Regular supercryl): 50%.

Shrinkage after dyeing and coning is 12%.

$$\text{Ring frame count } (Ne) = \frac{32}{1.69 \times 0.88} = 21.5s.$$

i.e. $21.5s \pm 0.2 = 21.3 - 21.7$.

5. Metric count 4/16(Hand knitting)

Mixing:

$8D \times 64$ mm (V-86) regular: 13%.

$5D \times 64$ mm (V-86) HB: 47%.

$3D \times 64$ mm HSSF (L2): 40%.

Shrinkage after dyeing and coning is 22%.

$$\text{Ring frame count } (Ne) = \frac{16}{1.69 \times 0.78} = 12.14s.$$

i.e. $12.14s \pm 0.2 = 11.94 - 12.34$.

6. Metric count 2/8(acrylic/nylon)

Mixing:

$3D \times 64$ mm Supercryl RSF: 45%.

$3D \times 64$ mm HSSF (L1): 40%.

22D × 80 mm Nylon: 15%.

Shrinkage after dyeing and coning is 12%.

$$\text{Ring frame count } (Ne) = \frac{28}{1.69 \times 0.88} = 5.4\text{s.}$$

i.e. 5.4s ± 0.2 = 5.2 − 5.6.

7. Metric count 2/24(Angora)

   Mixing:

   3D × 64 mm Supercryl RSF: 40%.

   3D × 64 mm HSSF (L2): 40%.

   22D × 80 mm Graylon (Nylon): 20%.

   Shrinkage after dyeing and coning is 12%.

$$\text{Ring frame count } (Ne) = \frac{24}{1.69 \times 0.88} = 16.1\text{s.}$$

   i.e.,16.1s ± 0.2 = 15.9 − 16.3.

8. Metric count 2/28(A/P)

   Mixing:

   3D × 64 mm IPCL (C): 40%.

   3D × 64 mm HLO Indarcryl: 40%.

   3D × 51 mm Polyester: 20%.

   Shrinkage after dyeing and coning is 22%.

$$\text{Ring frame count } (Ne) = \frac{28}{1.69 \times 0.78} = 21.2\text{s.}$$

   i.e., 21.2s ± 0.2 = 21.0 − 21.4.

9. Metric count 2/32(Soft wool)

   Mixing: 0.9D × 51 mm Acrylic Grey Pasupati Regular: 60%.

   0.9D × 51 mm Acrylic Grey H.B (PP): 40%.

   Shrinkage after dyeing and coning is 22%.

$$\text{Ring frame count } (N.E) = \frac{32}{1.69 \times 0.78} = 24.28\text{s.}$$

   i.e., 24.28s ± 0.2 = 24.08 − 24.48.

7.2   Types of instruments used for testing synthetic fibre parameters

Following are different instruments for testing different synthetic fibre parameters:

1. **VIBROJET 2000:** "Vibrojet 2000" is a combined automatic denier and tensile tester for all types of single fibres. Vibrojet 2000 tests single fibres for their fineness (denier) by loading them into a magazine, which is automatically put to the tensile tester by means of a robotic arm. More than 500 fibres can be stored in each magazine and be tested for their tensile properties automatically. In a batch mode, tests can be administered overnight without an operator. Test results include single results and statistics for denier, tenacity, elongation, young modulus, T/E curves and histograms.

2. **VIBROSKOP 500:** Vibroskop 500is an automatic instrument for the determination of the titer (dtex, den) of single fibre. By an advanced and patented approach of the vibration method (Vibroskop method) it assures best accuracy and reliability and it eliminates any influence of the operator.

3. **VIBRODYN 500:** Vibrodyn 500 is used for testing single fibre tensile properties. Vibrodyn 500 has been developed to cover the wide spectrum of requirements in tensile testing of single fibre. Programmable microelectronics guarantee maximum flexibility. The instrument is "one button operated" for easy and fast handling and also auto calibrated. Thereby, any operator influence is avoided, which means optimum accuracy and reliability of results. Vibrodyn 500 meets all international standards (ASTM, BISFA, ISO, DIN, …).

4. **VIBROCHROM 400:** Vibrochrom 400 is used for quick determination of whiteness. It is a flexible instrument for reliable and quick determination of whiteness, colour difference and fluorescence, which can be used for staple fibres and filament yarns.

5. **VIBROTEX 400:** Vibrotex 400 is used for easy determination of crimp properties of staple fibres such as crimp removal or contraction, crimp recovery and crimp stability. Results are represented graphically as well as in terms of figures on the connected PC.

7.3   Blending or mixing of man-made fibres

Mixing department is one of the most important departments for man-made fibre spinning. Proper blending of different component fibres, spraying of spin finish oils and water on fibres and proper conditioning of fibres have a direct impact on the process ability of the fibres in the manufacturing process. Mixing has been discussed in details in Chap. 2. Moisture regain of fibres influences the blend percentage of any blended yarn. When mixing two types of fibres, the quantity of fibres which are to be taken for mixing are to be calculated on oven dry weight (ODW) of fibres, otherwise blend percent will vary. It is more important when fibres with high moisture regain are used. Let us take an example for polyester/viscose (65/35) and polyester/wool (55/45) blended yarns.

First Case: Polyester/viscose (65/35),

We know MR% of polyester at 65% RH is 0.4%,

So, ODW of polyester = 65 kg − (65 kg × 0.4%) = 64.7 kg.

ODW of viscose = 35 kg − (35 kg × 13%) = 30.45 kg.

13% is the MR% of viscose fibres at 65% R.H.

Total weight of polyester and viscose based on ODW is equal to 95.15 kg.

So, polyester % on ODW basis = 64.7 × 100/95.15 = 68%.

Hence, if we add 65 kg of polyester in blend, we will get 68% polyester and rest is viscose percentage. So, we will add 3% less polyester in mixing i.e. 62% polyester and 38% viscose to get 65/35 P/V in the final yarn. Let us verify it:

Polyester = 62 kg − (62 × 0.4%) = 61.75 kg.

Viscose = 38 kg − (38 × 13%) = 33.06 kg.

Total = 94.81 kg.

Polyester % = 61.75 × 100/94.81 = 65.13%.

Viscose % = 33.06 × 100/94.81 = 34.87%.

Second case: Polyester/wool − 55/45.

Taking MR% of wool at 65% RH as 16% and that of polyester as 0.4%

Wool required on ODW basis = 45 kg − (45 × 16%) = 37.8 kg.

Polyester required on ODW basis = 55 kg − (55 × 0.4%) = 54.8 kg.

So, total weight of polyester and wool on ODW basis = 92.6.

So, polyester% on ODW basis = 54.8 × 100/92.6 = 59.18%.

Hence, if we add 55 kg of polyester in the blend, we will get 59.18% polyester and rest is wool percentage. So, we will add 4.18% less polyester in mixing i.e. 50.82% polyester and 49.18% wool to get 55/45 P/W in the final yarn. Let us verify it:

Polyester = 50.82 kg − (50.82 × 0.4%) = 50.62 kg.

Wool = 49.18 kg − (49.18 × 16%) = 14.32 kg/91.94 kg.

So, polyester % = 50.62 × 100/91.94 = 55%.

Wool % = 41.32 × 100/91.94 = 45%.

7.3.1   Blending of polyester/carded cotton in blow room mixing

Polyester = 52%, carded cotton = 48%.

Fibre taken: Polyester − 1.2D × 34 mm high tenacity fibres.

Carded cotton − carded sliver made of cotton (J34-RG/H4/S6 as per count spun).

Tint used for polyester = 40 g of tint/ton of polyester.

Total mixing = 3.5 tons.

Spin finish used on polyester:

LV40: 0.03%.

Water: 1%.

Procedures: At first polyester fibres are opened through bale opener and tint with spin finish and water is sprayed over opened fibres and collected in a bin being spread evenly on bin's floor.

Polyester fibres taken = 3.5 tons × 50% = 1.75 tons.

Carded sliver is opened in bale opener. At first 50 kg of polyester fibres are spread over the floor of a bin and above this layer 50 kg of opened carded slivers are sprayed evenly. One kg of water is sprayed over cotton. This is called the first layer. Likewise, 35 layers are made following above procedure. Then it is cut vertically and passed through the bale opener and evenly sprayed over floor in the bin this is called first toppling. Likewise, 2nd toppling is done. Then the mixing is finally ready for processing through blow-room lines. It is to be kept in mind that the final mixing should be kept for conditioning in mixing room for 24 h.

7.3.2   Blending of polyester (grey) and viscose (grey) in blow-room mixing (70/30)

Fibres taken:

Polyester = 1.4D × 44 mm grey recron – 70%.

Viscose = 1.5D × 44 mm grey Nagda viscose – 30%.

Spin finish = LV40 – 0.12%.

= 2152-P – 0.03%

= Water – 4%

Total mixing = 1 ton

Tint = 60 g/ton of polyester.

Procedures: Viscose bales are first hand opened and all chips, undrawn fibres are removed. Tuft sizes are reduced to 8–10 g, and kept for conditioning for 24 h. Then MR% of viscose and polyester fibres are measured. MR% of viscose and polyester fibres as found after conditioning are 13% and 0.4%, respectively.

To maintain polyester/viscose blend percentage 70/30, on ODW basis, 680 kg of polyester fibres and 320 kg of viscose fibres are to be taken.

68 kg of polyester fibre are first opened and spread over the floor of a bin. Then tint mixed with antistatic oil and water is spread evenly over the fibres. Over this 32 kg of hand opened viscose fibres are spread. Like this way, 10 such layers are produced. After completion of the mixing vertical cutting is done from the mixing and passed through the bale opener. This is the first toppling. Again layering is done & 2nd toppling is done through vertical cutting. Mixing is conditioned for 24 h before using in blow-room.

Note: Many mills avoid 2nd toppling and only one toppling is done. Conditioning for 24 h is also avoided. But this practice is not advisable. Problems like shade variation, differential dye take up may occur, if it is a grey yarn.

7.3.3   Polyester/cotton mixing for export

Polyester – 65%, combed cotton – 35%.

Fibre taken: Polyester – 1.0D × 32 mm high tenacity recron polyester grey.

Combed cotton – Taken from comber in sliver form.

Total mixing – 1 ton.

Spin finish used: LV-40 – 0.02%.

Water – 3%.

Procedures: Taking MR% of cotton as 8% and that of polyester as 0.4% at 65% RH,

ODW of polyester = 65 kg − (65 kg × 0.4%) = 64.80 kg.

ODW of combed cotton = 35 kg − (35 kg × 8%) = 32.2.

Total mixing based on ODW = 97.0 kg.

Now, 97 kg mixing contain 65 kg polyester fibre.

100 kg mixing contain 65 × 100/97 = 67.0%.

So, if we add 65 kg of polyester fibres in the mixing then we will get 67.0% of polyester for 100 kg mixing and rest 33% combed cotton. So, we should go for 2% or 2.0 kg less polyester for 100 kg mixing.

So, based on ODW, 63 kg of polyester and 37 kg of combed cotton should be taken for 100 kg of mixing to achieve 65/35 polyester/cotton.

At first, 30 kg of polyester fibre are taken from the bales, passed through blender and sprayed over the floor of a bin. Over it LV-40 (.02%) and water (3%) is evenly spread. Again, 33.0 kg polyester fibre are taken from bales, passed through bale opener and spread over previously opened and sprayed polyester fibres. Now, 37 kg of combed cotton sliver are passed through bale opener (+5 gauge) so that slivers are fully opened. These 37 kg of opened slivers are now manually spread over 63 kg of polyester fibre evenly. For 1 ton mixing, 10 such layers of 100 kg each are prepared one above another. Now, layers are vertically cut and passed through blender. This is called first toppling. First toppled fibre is heaped in a stack mixing manner in a bin. Now, 2nd toppling is done and the final mixing is kept for 24h conditioning.

Note: For export quality, tint is avoided.

7.3.4   Mixing of viscose melange fibres at blow-room stage

Types of mixing:

90% Viscose dyed + 10% viscose grey.

50% Viscose dyed + 50% viscose grey.

10% Viscose dyed + 90% viscose grey.

25% Viscose dyed + 75% viscose grey.

Fibre taken: Viscose dope dyed fibres − 1.5D × 44 mm Nagda.

Viscose grey − 1.5D × 44 mm Nagda Grey.

Spin finish: Elen 40 – 0.2%.

Water – 2%

Procedure: At first, dope dyed viscose (usually black) and grey viscose fibres are hand opened and passed through blender with +20 gauge separately. Then spin finish is sprayed on both grey and dyed fibres separately and conditioned for 24 h.

Dyed and grey fibres are weighed as per percentage. For 90% dyed melange 90 kg of dyed fibres are passed through blender with +5 gauge and spread over the floor of a bin. Then 10 kg of grey viscose fibres are passed through the blender with same gauge and evenly spread over dyed fibres. This process is carried out for the total mixing taken. Then vertical cuttings are done and passed through the blender for toppling. Toppling should be done two times.

7.3.5   Special precaution while running micro denier polyester – 0.8 denier

1. If dyed, then fibres are to be checked manually to sort out entangled fibres or any rope formation which are formed during dyeing.
2. Dyed fibres should be given one opening through blender.
3. Toppling in mixing should be made two times.
4. Carding hank should be 0.180–0.185.
5. Recommended speed: Cylinder – 390–400 rpm/Taker-in – 900 rpm/ Doffer – 14 rpm/Flat – 11.0 inches/min.
6. Flat waste % should be increased to maximum possible extent.
7. 3-Passage of drawing should be used.

7.3.6   Special precaution while running 100% viscose in rainy season

Fibres: Grey or dyed viscose – 100%

Special care:
1. Water spraying in mixing should be 1.5% maximum.
2. Heating lamps in blow room to be provided:
   (a) Over feed lattice.
   (b) Over calendar roller.
3. Heating lamps in drawing to be fixed over feed table.
4. Heating lamps in simplex to be set:
   (a) Over front roller.
   (b) Over creel.
5. Roving tpi should be on lower side.
6. Roving hank should be on finer side.
7. Ring frame cots shore hardness should be 65.
8. Recommended auto corner speed is 900 ypm.

After adjustment of the heating bulbs in blow-room and drawing, MR% of fibres from laps and finisher drawing slivers should be measured. Recommended MR% of fibres:

From laps: 12-13%.

From drawing sliver: 10 -11%.

Also, the following things are to be maintained:
- Calendar rollers pressure in bow room to be reduced.

- Beater speed in blow room to be reduced.
- Feed plate to Liken in gauge in card to be made wider.

Most important factor is the conditioning of viscose fibres for 24 h after opening from bales. This helps in better run ability as this increases the fibre relaxation.

7.3.7   How to alter blend percentage in mixing to some other blend% by adding only one component

Suppose, we have a cotton/viscose blend in ratio of 68/32 lying in the mixing bin. How to alter this blend% to 60/40 only by adding viscose, if asked by the management?

Let the total mixing be 100 kg.

i.e.

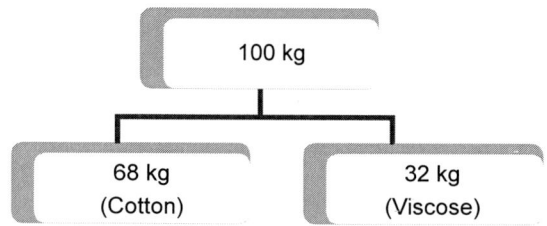

Then, $68 : x :: 60 : 40$.

or $x = 68 \times 40/60 = 45.3$ kg.

Where, x is amount of viscose in kg.

So, an extra amount of viscose to the extent of $(45.3 - 32)$ kg i.e. 13.3 kg are to be added to the 100 kg of mixing of 68/32 C/V to get 60/40 C/V.

Total mixing will be 113.3 kg.

Cotton% = $68 \times 100/113.3 = 60\%$.

Viscose% = $45.3 \times 100/113.3 = 40\%$.

7.3.8   How to calculate drawing hank in 1st passage of drawing to maintain a certain blend%

Suppose, we are to make polyester/combed cotton blend in 33/67 ratio on drawing. We need to know what drawing hank we should keep in the 1st passage of drawing for polyester and combed cotton so that in 2nd passage of drawing we can have our targeted blend percent being mixed with certain proportion of drawing can in feeding creel.

For doing such calculation, we should fix-up beforehand the following things:

(i)   2nd passage drawing hank.

(ii) No. of drawing can of polyester and combed cotton to be creeled in 2nd passage of drawing.

For the above case:

Let 2nd passage drawing hank = 0.134.

Polyester can creeled = 3 number.

Combed cotton creeled = 5 number.

So, in 1st passage, drawing hank of polyester = 62.16 grain × 33% × 8/3.

= 20.51 × 8/3 grain.

= 54.7 grain = 0.152 hank.

Note: 0.134 is the 2nd passage hank = 62.16 grain.

(Out of 8 cans, 3 cans of polyester added.)

For combed cotton, 1st passage drawing hank $= 62.16 \times 67\% \times \dfrac{8}{5} = 0.125$.

i.e. 1st passage polyester hank should be –0.152.

i.e. 1st passage combed cotton should be –0.125.

g/yd for polyester for 3 ends $= \dfrac{0.54}{152} \times 3 = 3.55 \times 3 = 10.65$.

g/yd for combed cotton for 5 ends $= \dfrac{0.54}{125} \times 5 = 4.32 \times 5 = 21.6$.

Polyester % $= \dfrac{10.65 \times 3}{32.25} = 33\%$.

Combed cotton % $= \dfrac{21.60 \times 5}{32.25} = 67\%$.

To find out hank of 1st drawing slivers for manufacturing blended yarn of certain blend ratio from mixing containing sliver waste of different blend ratio.

Suppose, we have a stock of polyester/cotton sliver waste with 33% polyester and 67% cotton, now, we are to make a mixing of polyester/cotton in the ratio of 45/55 by adding fresh polyester fibres with the sliver waste of P/C (33/67). We need to find out hank to be kept in the first passage of drawing for polyester and cotton so that we can get our desired blend percentage in 2nd passage of drawing.

Firstly we have to fix-up following two things before calculation:

(i)   Drawing hank to be produced in 2nd passage of drawing. Here it is taken as 0.134(62.19 grains/yd).

(ii) How many number of drawing cans of each component to be creeled in 2nd passage drawing. Here, it is 4 cans polyester and cotton each.

Total mixing (polyester and P/C waste):

800 kg fresh polyester (say) +10% waste of P/C sliver (33/67) = 880 kg.

In 880 kg total mixing, polyester content = 800 kg + 33% 80 kg or 27 kg (approx.) = 827 kg.

Cotton content = 67% of 80 kg x = 53 kg (approx.).

i.e. polyester – 94% , cotton – 6%.

In 2nd passage of drawing, percentage of sliver with polyester mixing for keeping 45% of polyester in the 2nd drawing sliver = 45%/.94 = 47.87%.

Percentage of cotton in 2nd drawing sliver = 52.13 from 100% cotton sliver + 47.87% of 6% cotton in polyester mixing sliver or 2.87% cotton = 55%.

1st passage drawing hank:

Polyester mixing – (62.19 × 47.87% × 8/4) or 59.54 grains/yd or 0.140 hank.

Cotton – (62.19 × 52.13% × 8/4) or 64.84 grains/yd or 0.129 hank.

7.4   Spin finish

A spin finish is a liquid or solid composition that is applied to the surfaces of man-made fibres in order to improve the processing of such fibres in short staple or long-staple spinning. The primary function of a spin finish is to eliminate the build-up of static electric charges on fibres during processing. This is achieved in two ways. First, the finish makes the fibre hydrophilic in order to facilitate charge dissipation (leakage). Second, it reduces the static and dynamic friction of the fibres, and subsequently the yarns, while they are moving in contact with machine parts, diminishing the generation and build-up of charge. Spin finishes can also control the amount of friction during processing. As an example, yarns experience drag as they pass over a ceramic guide or pass through the traveller during ring spinning. If the drag is too great, due to the degree of friction between fibres and machinery, fibres can be damaged. A spin finish can reduce the amount of friction to a level which avoids problems such as end-breaks in fibres. Friction can also cause local fusion of fibres, especially at points where the fibres rub guides and other machine parts during high-speed winding [7]. The effects of friction on textile fibres are discussed in the book by Gupta [8]. The effects of lubricants in spin finishes on reducing friction are discussed by Kutsenko and Theyson [9].

7.4.1   Types and application of spin finishes

The type of spin finish to be used depends on many factors [10, 11]. These include:
- The chemical properties of the fibres.
- The type of spinning process, e.g. short- or long-staple spinning, or continuous filament spinning.
- The technique used for applying the finish.

The fibre producer can often give the best advice for selecting the spin finish. It is recommended that the spinning mill applies the fibre manufacturer's own spin finish so that it can be closely matched with the fibre.

7.4.2   Key requirements for spin finishes

Spin finishes must meet a range of requirements if they are to be effective and to add value to the final product [12]. These requirements can be classified into three main themes:
- interaction with fibres,
- interaction with processing conditions and machinery,
- safety and other issues.

It is essential that the application of spin finishes should not affect fibre morphology and quality. A spin finish must be readily adsorbed and have good adherence to the fibre surface. It must have the ability to wet the fibre surface and spread evenly over it to avoid dry friction. On the other hand, it must be easily removed before dyeing [13]. It must also have no effects on dyeing absorption, i.e. the dye affinity of fibres, or on dye fastness. Some spin finishes have been used to block certain dyes where this is required [14]. It is important to ensure that the finish does not affect the shelf-life of the final yarn. Some finishes use titanium dioxide to enhance colours in fabric but the additive is abrasive and can weaken the yarn [7].

Spin finishes must be able to withstand processing conditions. This requires characteristics such as:
- A high resistance to intensive rubbing, i.e. low wear ability, and resistance to flaking when subjected to friction,
- Thermal stability and reasonable ability to undergo evaporation (sublimation) with no ill-effects,
- A constant viscosity with increase in temperature,
- Lack of foaming.

Spin finishes must not cause damage to metallic and non-metallic surfaces of machine parts (e.g. aprons, cots of top rollers of the drafting system, rotors in spinning machines).

More generally, spin finishes must be toxicologically and physiologically safe [15].

7.4.3   Suggested application of spin finish percentage on synthetic fibres at mixing stage

Application of different types and percentage of spin finish for polyester, viscose and acrylic fibres and their blends are given in Table 7.8.

Table 7.8: Application of spin finish

Sl. no.	Fibres	Spin finish and water percentage (owf)
1.	Acrylic grey coarser denier	Wet spun: V SAS200 –0.28% Water – 8% Dry spun: V SAS200 – 0.45% 5280 – 0.06% Water – 11%
2.	Acrylic grey (1.5D)	VSAS 200 – 0.17% 5280 – 0.06% Water – 7%
3.	Acrylic grey (0.9D)	Elen 40 – 0.07% 5200 – 0.07% 092 – 0.02% Water – 6.0%
4.	Polyester grey – 1.2D/1.4D	LV40 – 0.02% 092 – 0.02% 2152(P) – 0.04% Water – 4%
5.	Polyester dyed/grey – 2.0D	Elen 40 – 0.02% 092 – 0.02% Water – 5.0%
6.	Polyester grey – 0.8D/1.0D	VSAS-200 – 0.06% Water – 3.0%
7.	Viscose grey/dope dyed – 100% (1.5D)	Elen 40 – 0.2% 5200 – 0.1% Water – 2%
8.	Polyester/viscose blend (65/35 or 70/30	LV – 40 – 0.30% 2152 (P) – 0.10% 0.92 – 0.02% Water – 3%

For TBL polyester in mixing, water % to be increased as:

- 40% TBL in mixing: Water – 3.5%.
- 65% TBL in mixing: Water – 4.0%.

Note: Percentage of spin finish and water which is to be applied on fibres at mixing stage depends on outside RH condition. Its percentage varies in winter, rainy and summer season. In extreme climatic condition, this variation is more pronounced.

7.5   Running of own dyed polyester fibres in spinning

There is a very good demand of 100% dyed polyester yarns in the market in the count range of 15s–30s. But spinning of 100% dyed polyester fibres is very difficult due to various types of problem arising right from carding up to cone winding as follows:

- Coiler tube choking in carding and drawing,
- Cylinder loading & excess flies liberation in cards,
- Lapping on top and bottom rollers in drawing, simplex and ring frame.
- Excessive cuts in winding.

To overcome the above problems to a manageable extent, the following conditions are to be maintained:

1. Quality of water used for dyeing polyester fibres must be soft and is to be tested to adhere to the norms which is shown here.
2. Proper cleaning and washing of dyed fibres after dyeing is needed.
3. Proper RC (reduction clearance) after dyeing is a must.
4. Two times RC is preferred for dark and heavy dark shades.
5. Choice of denier for own dyed fibres are important and it is always better to choose coarser deniers.

Deniers should be so chosen that number of fibres in yarn cross-section of remain between 70 and 75.

7.5.1   Dyeing of polyester fibres

The following dyeing process can be used for polyesters fibre dyeing to get a good result in spinning process.

Machine used: HT/HP, M : L = 1: 4.

1st Step: fibres are washed at 70 °C for 10 min and water is drained out.

2nd Step: Bath temperature is set at 50 °C and some levelling cum dispersing agent (0.5 g/l) is added. Then, acetic acid (0.60 g/l) is added to keep the pH of the bath in between 4.5 and 5.0. Pre-dissolved disperse dye stuff is then added into the vessel. The temperature adjustments are done as follows:

1. Raise the temp. from 50 to 90 °C @ 1.5–2.0 °C/min.
2. Raise the temp. from 90 to 130 °C @ 1.0 °C/min.
3. Keep the temp. for 30–45 min.
4. Drain the bath at 130 °C.

3rd Step: For reduction clearance, the following process to be followed:

1. Hot wash in boiling temp. for 30 min and drain the water.
2. R/C with the following chemicals:

    Non-ionic detergent – 0.25 g/l

    Hydro – 1.5 g/l

    NaOH – 1.50 g/l

    Dispersing agent – 1.2 g/l

    (Sera cone PU –Dye star Co.)

    Temp – 85 °C

    Time – 30 min

    For dark and heavy dark shades, R/C is to be carried out for second time following the above procedure after draining out the chemical.
3. Drain out chemicals and hot wash at boiling temp. for 15 min.
4. Neutralisation with acetic acid @ 0.2 g/l for 15 min at 35 °C.
5. For finishing – Use antistatic agent @ 0.5–0.7% at 50 °C for 20 min (antistatic oil – Sap cost at F as RSA54) and take out the fibres without draining the bath. pH of the bath is to be kept in between 4 and 5.

Then the goods are hydro extracted so that moisture after hydro extraction comes to 8–9%.

Prior to drying, goods are opened in an opener machine and then dried in a dryer at 90 °C so that moisture in the fibres remain in between 4% and 5%.

Note: R/C process is meant for removing oligomer from the fibres surfaces. This can be better achieved through the use of multipurpose oligomer remover (SaraconP-Nu, marketed by Dye star Co.) which removes oligomer residue from dying machines and fibre's surfaces.

7.5.2   Checking dyes applied on polyester fibre for tinting

Sometimes the tint applied on polyester fibres do not wash out either by cold or hot water washing. If possible, tinting process should be avoided. However, due to practical reasons, it is not possible always to do so. However, wash ability of tint should be checked by QA department before applying on fibres. Checking procedure is given below:

Fibres taken –100 g

Dissolve 10 mg (0.01%) of tint in 4 g of water (4%). This tinted solution is evenly sprayed on fibres, toppled and kept for 8 h at room temp.

For cold wash : Tinted fibres are kept fully immersed in ordinary water for 30 min. Then wash in the running water with hands rubbing. The tinted colour should get washed away completely.

For hot wash: Tinted fibres are taken in a beaker filled with water so that fibres are fully immersed in the water. Then heat the beaker at 60 °C and keep for 30 min. The tinted colour should get washed completely.

If slight presence of tinted colour is found either after cold or hot wash, then its wash ability is poor and that particular dye stuff should be discarded to be used as a tint.

7.5.3   Water quality requirements for dyeing

It is very much important to maintain proper quality of water used in dyeing. The quality requirement of water for dyeing of polyester fibres is given in Table 7.9.

Table 7.9: Water quality

Sl. no.	Test parameters	Test method	Unit	Water quality requirements
1.	Colour	IS201:1992	Hz unit	Max. 5
2.	Turbidity	IS 3025	NTU	Max. 2
3.	pH value	IS 3025	–	6.5–8.5
4.	T. Alkalinity (as $CaCo_3$)	IS 3025	mg/l	Max. 150
5.	T. Hardness (as $CaCo_3$)	IS 3025	mg/l	Max. 2
6.	Chloride (as Cl)	IS 3025	mg/l	Max. 100
7.	Sulphate (as SO_4)	IS 3025	mg/l	Max. 100
8.	Aluminium (as Al)	IS 3025	mg/l	Max. 0.1
9.	Iron (as Fe)	IS 3025	mg/l	Max. 0.1
10.	Manganese (as Mn)	IS 3025	mg/l	Max. 0.1

7.6   How to utilise soft waste which are generated during yarn manufacturing

Good amount of soft and usable wastes are always generated during yarn manufacturing process in synthetic spinning mills (dyed P/V, dyed acrylic, dyed viscose). All these waste can be used to make yarn of 20s Ne in 3-colour categories, viz. brown, blue and black. Jogging and track suits may be manufactured out of these yarns. Utilisation of waste is made as follows:

Composition:
- Flat strips (15%),
- Bonda (60%),
- Roving (5%),
- Fresh fibres (20%, mainly polyester).

Procedure

Flat strips, rovings and fresh polyester fibres are opened through blender separately and individually. Then all these three types of opened fibres are made to pass through blender and this mixing is kept separately. Bonda wastes are also opened through blender. Now 40 kg of (flat + roving + fresh) fibres and 60 kg of bonda are passed through blender. Layering is done in such a way that 40 kg of (flat + roving + fresh) fibres are first spread over the floor of a bin and above it, 60 kg of bonda is spread. Like this way, 1 ton mixing is done. Then it is vertically cut and toppling is done through blender. One time toppling is enough. Different categories of wastes are maintained under the list of brown, blue and black are shown below in Table 7.10.

Table 7.10: Different categories of wastes

Brown category	Blue category	Black category
Coffee brown	Navy blue	Black
Brown	Royal blue	Grey black
Cadbury	Air force blue	Charcoal
Fawn	Medium blue	Pista
Camel	Green	Medium black
Khaki	Bottle green	Light grey
Chocolate brown	Parrot green	Ash grey
Beige	Sky blue	Mouse
Red	Slate grey	
Maroon		

Contd...

Contd...

Brown category	Blue category	Black category
Mustard		
Rust		
Olive		
Mehandi		
Silky		
Violet		

7.7   Blend % checking from each lot from card's sliver

Frequency: From each lot. Four reading are taken.

Standard:

    (i)   Range out of 4 reading – 1.5–1.7 units, (Avg. ±1unit) – for grey.

    (ii) Range out of 4 reading – 1.8–2.0 units (Avg. ±1.5 units) for dyed.

How to check: Card sliver is taken and four numbers of samples weighing 1 g approx. are taken from different positions of the card sliver and chemical analysis is done for blend% testing. If any lot shows any higher deviation than standard value, then testing is again done from drawing sliver. If same observation persists, then production department and sales department should be informed. If the lot is a grey one, then the variation problem is serious, but if it is dyed, then the problem is not so serious, but one should be very careful about blend variation whether it is grey or dyed.

In-process (carding): Working of own dyed polyester fibres are to be watched for each lot in carding section. Problems like cylinder loading. Coiler tube choking, excess flies liberations are occasionally observed. If any dyed lot is found giving problem in carding, dye/house head should be informed immediately. This problematic lot should be under QA's observation upto final winding stage. A joint meeting among QA, production and dye house should be held so that this problem can be minimised or avoided in the next time when this same lot comes.

In-process checking for drawing, simplex, ring frame and outgoing checking are almost alike cotton.

Shade matching checking: This is one of the most important checking for dyed lots. Shade matching registers should be maintained for dyes used for p/v, polyester, polyester/acrylic, acrylic (all own dyed fibres) separately. As soon as yarns are spun is ring frame, one lea should be made and properly pasted in that particular register. When this same lot comes for manufacturing, the same above process is to be followed and should be matched with the previous one. If the shade matches, then there is no problem. If it does not match then immediate information is to be given to Dyeing Master. Sales department is to be informed also. This shade matching procedure or system should be based on previous lot and not on the first lot. To make it clear, let us consider that one particular dyed lot was repeated four times. Next shade matching should be done on fourth one, and not on first one.

7.7.1   To prepare chemical solution for confirming blend % of synthetic blended yarn

To dissolve all types of fibres like cotton, viscose, wool, acrylic and nylon except polyester fibres:

How to prepare the solution: Take 7cc of distilled water and add it to 19cc, H2SO4 (laboratory reagent). Take 20cc of glacial acetic acid and pour it slowly into the previous solution containing water and cone H2SO4. Allow the solution to cool down. For increasing the volume of the solution, it can be taken in multiple of 7cc–19cc–20cc.

For chemical analysis of any blended yarn, first measure the moisture regain percentage of yarns. For this purpose, cut the yarn sample of about 1 g into small pieces of 1″ cut length roughly. Further weigh and put it on watch glass. Keep the yarns sample in a drying or conditioning oven where temperature is maintained in between105 and110 °C. Keep the sample for 2½ h and then transfer the sample in a stoppered glass bottle for weighing and measuring moisture regain percentage.

Then the sample is transferred in a glass beaker containing the above solution. Chemical solution taken should be sufficient to cover up all the small pieces of yarns. Keep this beaker in a drying oven for 10–15 min and then take out the beaker. Stir the solution with yarn gently with a glass rod for 2.3 min. Take out the chemical solution from the beaker taking care not to lose any fibre.

Washing is a very important part of chemical analysis. Distilled water is to be poured with glass rod. Water is to be drained out and repeat this process again. Then take roughly 25–30cc distilled water again in the beaker and pour roughly 5cc liquid ammonia into it for neutralising the acid present in the yarn. Again wash the samples thoroughly, with cold distilled water following the above procedure. Then squeeze the sample by placing the sample on palm and give pressure on the sample by the thumb of the other hand. Transfer the sample to take the oven dry weight in a conditioning oven. Calculate the blend percentage based on oven dry weight of the sample.

• **To dissolve polyester fibres**

Preparation of the reagent: Take 25 g of trichloroacetic acid in 100 ml of methylene chloride at room temperature. Dip 0.5 g (over dry weight) of the yarn sample into 25 ml of the reagent as prepared above and keep for 5 min at room temperature. After subsequent rinsing with methylene chloride and later with acetone, dry the residue in oven and weigh.

• **To dissolve viscose fibres**

Viscose fibres are dissolved in 60% H_2SO_4.

Preparation of 60% H_2SO_4: Add 360cc of conc. H_2SO_4 to 390cc of distilled water and standardise the strength to 102° TW. Cool down the solution to normal temperature at room temperature.

Twaddle meter specification: Twaddle-meter no-5, Range 96–120°.

Always add acid to water (Remember ATW).

• **To dissolve wool fibres**

Wool fibres are dissolved in 5% caustic soda solution at boil.

Preparation of 5% caustic soda solution: Take 5 g of pure NaOH pellets (laboratory reagent) into 100cc of distilled water. Heat it to boil for making a clean solution.

• **To dissolve acrylic fibres**

Acrylic fibres are dissolved in n,n-dimethyl formamide (DMF) solution at room temperature. n,n-Dimethyl formamide solution (laboratory reagent) is available in the market in pure form.

• **To dissolve nylon fibres**

Nylon fibres are dissolved in cone – formic acid.

7.7.2   Precautions to be taken for chemical analysis

(i)   The sample for analysis should be taken on over dry weight.

(ii) The temperature of the conditioning oven should be kept in between 105 and 110 °C. This temperature is to be checked from time to time.

(iii) Yarn under chemical analysis should be cut into small pieces before taking MR%.

(iv) Stoppered weighing bottle is to be used for sample weighing and testing.

(v)   Oven dry weight sample should never be exposed in open atmosphere. It should be quickly transferred to the stoppered bottle from the heating chamber by tweezers, not by finger.

(vi)   Chemicals are to be prepared freshly before testing.

(vii)   Washing of the chemically treated fibres should be done with proper care for acid or alkali neutralisation. Checking by pH value paper is preferred which should be between 7 and 8.

7.8   QA checking on process parameters and other in synthetic section

This checking system is almost alike cotton process except on certain areas which are given here in Table 7.11. Bottom fluted rollers gauge (C–C) and corresponding saddle gauge in simplex and ringframe is detailed in Table 7.12. Table 7.13 gives key features of different fibres characteristics and apparel application.

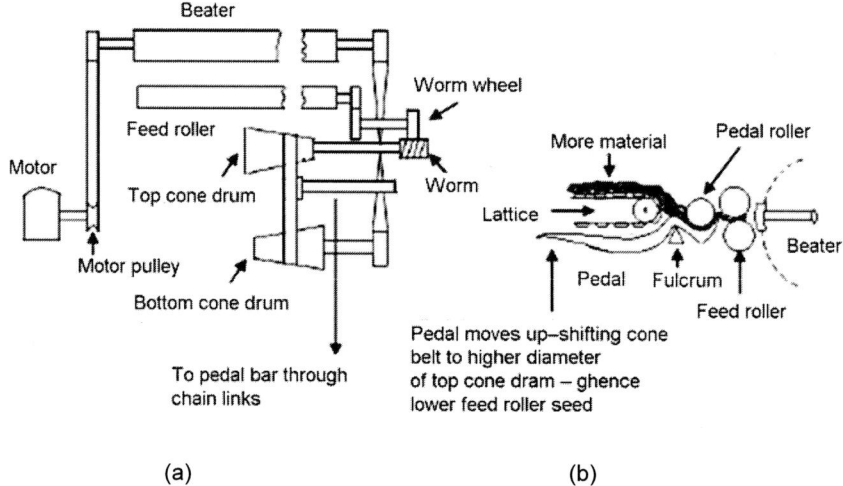

(a)   (b)

Figure 7.1: Feed control system in blow room (A) speed control by cone drums and (B) piano feed mechanism

Table 7.11: Checking

Incoming (raw material	This checking is very much, limited unlike cotton. For synthetic fibres bales, 5% of the total bales consignment arrived in the mills, to be checked for (i) Denier, (ii) Length, (iii) Over length or uncut fibre, (iv) Undrawn fibres and chips. Difficult to check the denier of micro denier fibres.

Contd...

Contd...

In process (blow-room)	Lap rejection checking in each shift for 1 h max. lap rejection which can be tolerated is 5%. If it is more than this, stop the line and inform the production department. Rejected laps are all sent back for refeeding which deteriorates the quality of the yarn.
	Piano feed mechanism and cone drum belt movement checking frequency: Once in a day (Fig. 7.1).
	How to check: for proper belt movement checking, take out loose fibres sheet coming out of reserve box and look at the belt movement. In this case, feed roller speed will go up and belt will move towards driver cone drum side. Continue to take out from time to time in a gap till the full lap is formed. Take out roughly 5–6% of the total lap weight. If the lap weight comes within the tolerance limit, then the piano feed mechanism is OK. Similarly, go on adding the same amount of fibres and check the belt movement on cone drum which should be on bigger diameter of driven side. Check the full lap weight after completion.
	Yard to yard full lap CV% after mixing change. Full lap CV% should be 1.2–1.3. This process is applied only for lap feed card.
In process (carding section)	Neps/100 sq. inch in cards web.
	Frequency –Daily once from all cards.
	Standard: max. 4 neps/100sq. inch.
	How to check: A wooden board of 100 sq. inch in total area is to be made. One handle is to be attached with the board for proper holding by hand. Now the card-web coming out of Doffer is to be checked one time from three different positions as LHS/Middle/RHS. Neps and unopened fibres are to be counted by naked eyes and average of three sides is to be noted.

Table 7.12: Bottom fluted rollers gauge (C–C) and corresponding saddle gauge in simplex and ringframe

Simplex		
Fibre length	Front–back (C–C) mm	Saddle gauge
44 mm	70–70 mm	LF-1400
51 mm	70–74 mm	Front (C–C) mm + 8 mm Back (C–C) mm − 5 mm
64 mm	73–80 mm	LF-1400A
76 mm	80–80 mm	Front (C–C) mm + 7 mm Back (C–C) mm −5 mm
Cotton	42–52 mm	
Ring frame		
44 mm	72–70 mm	Front – 78 mm, Back – 70 mm
51 mm	72–70 mm	Front – 78 mm, Back – 70 mm
64 mm	72–70 mm	Front – 78 mm, Back – 72 mm
76 mm (with apron recess roller)	72–65 mm	Front – 78 mm, Back – 60 mm
Cotton	42.5–60 mm	Front – 50 mm, Back – 60 mm

Table 7.13: Fibres characteristics and apparel application

Fibres	Key important characteristics	Versatility and process ability	Complementarily with other fibres	Main apparel application
Cotton	1. Breathable 2. Hydrophilic 3. Crisp 4. Dry handle 5. Low lustre	Medium	Medium	1. Shirting 2. T-shirts 3. Jeans wear
Polyester	1. High lustre 2. Hydrophobic 3. Hard wearing 4. Crease resistant	High	High	Found in 100% form and blended across most segment
Acrylic	1. Soft 2. Bright 3. Stretchy	Low	Medium	1. Hosiery 2. Sports wear 3. Outdoor wear
Viscose	1. Excellent drape And dye take-up	Low	Medium	Ladies top weight especially print
Wool	1. Extensible 2. Crease resistant 3. Non-flammable 4. Wear	Low	Poor	1. Knit wear 2. Formal woven

7.9   References

1. Chao, Nelson Ping Ching, Blending cotton and polyester fibers –Effects of processing methods on fiber distribution and yarn properties, M.Sc. thesis, Georgia Institute of Technology, September, 1963.

2. Speakman, J.B., Fibre Blends, J.Text. Inst. (Proc.). Oct. 1958, Vol. 49, Issue 10, p P580–P583.

3. Baykal, Pinar Duru, Babaaslan, Osman, Erol, Rizvan, Prediction of strength and elongation properties of cotton/polyester-blended OE rotor yarns, Fibers Text. East. Eur. January/March 2006, Vol. 14, No. 1 (55).

4. Cyniak, Danuta, Czekalski, Jerzy, Jackowski, Tadeusz, Quality analysis of cotton/polyester yarn blends spun with the use of a rotor spinning frame, Fibers Text. East. Eur. July/September 2006, Vol. 14, No. 3 (57).

5. Sevkan, Ayse, Kadoglu, Husejin, An investigation of ring and open-end spinning of flax/cotton blends,Tekst. Konfeksiyon, 3/2012, Vol. 22, 218–222.

6. Hegde, Raghavendra R., Dahiya, Atul, Kamath, M.G., Cotton Fibers, Updated: April, 2004.

7. Lord, P.R., 2003, Handbook of Yarn Production, Woodhead Publishing, Cambridge.

8. Gupta, B.S. (ed.), 2008, Friction in Textiles, Woodhead Publishing Limited, Cambridge

9. Kutsenko, M. and Theyson, T., 2008, Finishes and Treatments to Control Friction in Textile Fibres, in B.S. Gupta (ed.), Friction in Textiles, Woodhead Publishing, Cambridge.

10. Oldrich, P., Reichstadter, B., 1979, Processing of Polyester Fibres, Elsevier, Amsterdam.

11. Hawary, I.A., 2008.Technology of Artificial Fibres Processing, Part I: Short Staple Processing,Lectures, TED, Alexandria University, Egypt.

12. Klein, W., 1994, Man-Made Fibres and Their Processing, Stephen Austin & Sons, Manchester.

13. Bernholz, W.F., Redston, J.P., Schlatter, C., 1974, Spin finish usage and compounding for man-made fibres, Polymer-Plastics Technology and Engineering, 3(1), 1–27. doi:10.1080/03602554085020.

14. Selivansky, D., Walsh, W.K., Frushour, B.G., 1990, Interactions of spin finishes with partially oriented poly(ethylene terephthalate) fibers: Part 1: The nature of finish/fiber interactions, Text. Res. J., 60, 33–41. doi: 10.1177/004051759006000105.

15. Mizerovskii, L.N., 2000, Improved spin finish for synthetic fibres: A new approach to solving the problem, Fibre Chem., 32(3), 221–260.

8.1 Production calculation in spinning mills

8.1.1 Spinning preparatory

1. Blow-room production per 8 h in kg at 100% efficiency

$$= \frac{D \times R}{\text{Lap hank}} \times \frac{3.14 \times 60 \times 8}{36 \times 840 \times 2.2}$$

$$= \frac{D \times R}{\text{Lap hank}} \times \text{Constant } (0.022),$$

Where, D = bottom calendar roller diameters in inches,
R = bottom calendar roller rpm,

$$\text{Lap hank} = \frac{0.54}{\text{Lap wt (g/yd)}}$$

2. Cards delivery rate in meter per minute (MPM)

$$= \frac{\text{Dofter dia (inch)} \times 3.14 \times \text{Doffer rpm}}{36 \times 1.09} \times \text{Tension draft}$$

Cards production per hour at 100% efficiency in kg

$$= \frac{\text{MPM}}{\text{Card hank}} \times \frac{100 \times 60}{2.54 \times 36 \times 840 \times 2.2}$$

$$= \frac{\text{MPM}}{\text{Card hank}} \times \text{Constant } (0.035)$$

3. Drawing and simplex production per hour at 100% efficiency in kg

$$= \frac{\text{Hank meter reading per hour} \times \text{No. of (delivery/spindle)}}{\text{(Sliver/roving) hank} \times 2.2}$$

4. Comber production in kg/h at 100% efficiency is given by:

$$\frac{G\left(1 - \frac{p}{100}\right)}{7 \times Z} \times N \text{ Where, } G = \text{g/m of ribbon lap,}$$

p = noil %, N = nips/min, Z = feed roll wheel.

Considering E 7/4 model comber:

$$G = 60 \text{ g/m},$$
$$p = 15.5\%,$$
$$N = 200,$$
$$Z = 74 \text{ T},$$

So, kg/h $= \dfrac{60\left(1 - \dfrac{15.5}{100}\right) \times 200}{7 \times 74} = 19.4$ kg/h at 100% efficiency.

8.1.2   Spinning

5. Ring frame production in terms of lbs/spindle/8 h at 100% efficiency

$$= \frac{3.14 \times 60 \times 8 \times \text{front roller rpm of 1" diameter}}{36 \times 840 \times \text{Count}}$$

$$= \frac{\text{Constant } (0.05) \times \text{front roller rpm of 1" diameter}}{\text{Count}}$$

Ring frame production in terms of g/spindle/8 h at 100% efficiency

$$= \frac{\text{Constant } (22.68) \times \text{front roller rpm of 1" diameter}}{\text{Count}}$$

Constant (0.05) can be used to calculate hanks delivered by the machine through the following relationship:

Hanks produced in 8 h at 100% efficiency = 0.05 × delivery roll diameter in inches × delivery roll rpm.

8.1.3   Post spinning

6. Auto winder production/drum/8 h at 100% efficiency in kg

$$= \frac{\text{Delivery speed in meter/min} \times 60 \times 8 \times 1.09}{840 \times \text{Count} \times 2.2}$$

$$= \frac{\text{Delivery speed in meter/min} \times \text{Constant } (0.28)}{\text{Count}}$$

7. Manual winder PSM (PS Metler) with drum diameter = 3.0″
Production/drum/day (24 h) at 100% efficiency in kg

$$= \frac{314 \times 3 \times 60 \times 24 \times \text{Del drum rpm}}{36 \times 840 \times 2.2 \times \text{Count}}$$

$$= \frac{\text{Constant } (0.20) \times \text{Del drum rpm}}{\text{Count}}$$

8. Winder with delivery drum diameter of 3.75″

Production/drum/day (24 h) at 100% efficiency in kg

$$= \frac{\text{Constant } (0.254) \times \text{Del drum rpm}}{\text{Count}}$$

9. Ring doubling:

Production of different ring doubling machines are shown in Table 8.1.

Table 8.1: Production/day(24 h)/machine at 100% efficiency in kg (P)

Machine make	No. of spindle	Front roll diameter	Constant	P
Textool	400	2.00″	54.37	$\frac{54.37 \times \text{Actual FRS}}{\text{Count}}$
UMW and minerva	408	2.50″	68.3	$\frac{68.3 \times \text{Actual FRS}}{\text{Count}}$

Note: Actual FRS means front roller rpm being noted with actual diameter, not being converted to 1″ diameter.

10. TFO

Production/machine/day (24 h) at 100% efficiency

$$= \frac{\text{Procution constant} \times \text{Acutal FRS}}{\text{Count}}$$

Table 8.2 shows the production constant values for TFO machines of different makes.

Table 8.2: Production constants for TFO machines of different makes

Machine make	No. of spindle	FR diameter (inches)	Production constant
Vijay Luxmi	144	3″	29.36
Prenna	132	3.5″	31.40
Star-Volkman	120	4″	32.62
Volkman	168	4″	45.67

8.1.4   HOK

HOK is defined as total operative hours required to produce100 kg of yarn with a standard count of 40s Ne.

How to calculate HOK:

At first, we have to calculate the total hands involved up to either ring frame or packing stage. Then average count of the ring frame section is to be calculated and converted to 40s count and production is to be adjusted in the basis of 40s count.

Following example will make the things clear. Here HOK is calculated up to ring frame for a cotton spinning mills of 25,000 spindles. Hands engaged are as follows:

1. Blow-room: 4 hands/8 h (two number of lines with chute feed).
2. Carding: 3 hands/8 h (24 cards − 12 cards/line through chute).
3. Drawing and simplex: 17 hands/8 h (Drawing − 6 RSB and2 Do/6; Simplex − 8 LF1400A with 120 spindles/mc.
4. Ring frame: 37 hands/8 h (24 ring frame with 1008 spindles/mc). Average production/day − 9300 kg.

Average count − 34s (average count calculation will be discussed later on).

1. Ring frame:

   Total operative hours = 37 (hands) × 3 (shift) × 8(h) = 888.

   Production with count adjusted to 40scount = 9300 × 34s/40s
   = 7905 kg.

   HOK = 888 × 100/7905 = 11.23.

2. Simplex and draw frame:

   For 7905 kg yarn production we must go far 2% more production in simplex and drawing i.e. drawing/simplex production will be (7905 + 2% of 7905) kg or 8063 kg.

   Total operative hours = 17 × 3 × 8 = 408.

   HOK = 408 × 100/8063 = 5.1.

3. Carding:

   It is required to keep 5% more production in carding than draw frame to make a regular flow of materials to drawing and simplex for production balancing. Therefore, carding production should be (7905 + 7% of 7905) kg or 8458 kg.

   Total operative hours = 3 × 3 × 8 = 72.

   HOK = 72 × 100/8458 = 0.85.

4. Blow room:

   Blow room production should be − (7905 + 10% of 7905) kg
   = 8696 kg.

   Total operative hours = 4 × 3 × 8 =96.

   HOK = 96 × 100/8696 = 1.10.

Hence, to get 7905 kg of yarns production, we have added 10% more production from blow room to compensate for wastes for production balancing.

So, total HOK up to ring frame:

Blow room	–	1.10
Carding	–	0.85
Drawing/simplex	–	5.10
Ring frame	–	11.23
Total:		**18.28**

18.28 HOK is very high and no mills can survive with this high value of HOK. It should be 11–12 HOK up to ring frame. Automation has become nowadays very urgent for survival particularly where hands engagements are more. With the introduction of auto doffing in ring frame, hands required may be reduced by 12.0 hands (ring doffers) per shift. Hence, HOK in ring frame will come down to 7.59. Besides this, ring frame sides will have to be increased from 12 sides to 16 sides (One ring frame with 1008 spindles = 4 sides). For this purpose, ring frame working should be made much better in terms of breakage percentage. Maximum 2% breaks are to be permitted (2% means 2 breaks/100 spindles/h). Simplex bobbin size should be bigger so that creel changes come down.

The above task is not only be rested on production people but quality assurance people should be equally involved to achieve this target.

8.2   How to find out the average count in ring frame section

Average count in ring frame should be calculated through the following formula:

$$\text{Average count} = \frac{\Sigma \, (\text{Countwise production lbs} \times \text{Actual count})}{\text{Total production of all counts in lbs}}$$

Following is an example of calculating the average count of a spinning mill having 24 ring frames of 1008 spindles with the following counts:

20s Carded	–	4 ring frame
24s Carded	–	6 ring frame
20s Combed	–	8 ring frame
24s Combed	–	2 ring frame
30s Combed	–	4 ring frame
Total:		**24 ring frames**

g/spl/8 h:

    20s Carded – 232 g @ 88% efficiency

    24 s Carded – 186 g @ 90% efficiency

    20 s Combed – 265 g @ 88% efficiency

    24 s Combed – 209 g @ 90% efficiency

    30 s Combed – 176 g @ 94% efficiency

Countwise production with the above particulars is given in Table 8.3.

Table 8.3: Countwise production

Count	R/Fr allocation	g/spl/8 h	Prod/8 h(kg)	Prod/8 h (lbs)	Count ×prod. (lb)
20s Carded	4	232	935	2057	20 × 2057 = 41140
24s Carded	6	186	1125	2475	24 × 2475 = 59400
20s Combed	8	265	2137	4701	20 × 4701 = 94020
24s Combed	2	209	421	926	24 × 926 = 22224
30s Combed	4	176	710	1562	30 × 1562 = 46860
				Total = 11729	Total = 263644

$$\text{Average count} = \frac{263644}{11729} = 22.48s$$

Parta system

The parta system is widely used in Birla Companies as a performance measurement and control system. The 'parta' represents the commitment of the unit in charge to the chairman of the company or the group. It is arrived at after taking into account various factors like manufacturing capacity, operational efficiency, consumption norms, market situation, supply of materials and labour productivity[1].

Here, an example is given to show that spinning 1 count coarser than parta count can balance the profit even with lower efficiency than parta efficiency. Suppose, parta efficiency is 92% against spinning count 30s, but actual efficiency achieved is 90.57% against spinning count 29.05s. What will be its impact on profit?

Let us see:

Count spun – 30sK, g/spl/8 h – 176 g at 100% efficiency.

Ring frames engaged = 20 mcs with 1008 spls/mc.

Production/mc/day @ 90.57% efficiency = 482 kg.

Total 20 ring frames production @ 90.57% efficiency = 9640 kg.

Production/mc/day @ 92% efficiency = 490 kg.

Parta efficiency (A) – 92%.

Actual efficiency (B) – 90.57%.

Parta count (C) – 30s.

Actual count (D) – 29.05s.

Actual ring frame production (G) – 9640 kg.

Ring frame production converted to parta efficiency:

$$E = \frac{G \times A}{B} E = 9792 \text{ kg.}$$

Ring frame production on parta efficiency and parta count

$$F = \frac{E \times D}{C} F = 9482 \text{ kg.}$$

Conversion cost Rs. 35/kg.

Production gain for count being coarser $(E - F) = 310$ kg.

Money gain for spinning coarser count = 310 kg × Rs 35 = Rs 10,850.

Production loss due to low efficiency = 9640 – 9792 = (–)152 kg.

Money loss on parta efficiency = 152 kg × Rs. 35.0 = (–) Rs. 5320.

Conclusion: Money gain for coarser count = Rs. 10850.

Money (loss for) efficiency low = Rs. 5320.

Net money gain (+) = Rs. 5530.

Standard waste generation at different points in the spinning process is given in Table 8.4.

Table 8.4: Standard waste generation in a cotton spinning mill

	Carded	Combed
Blow room droppings	4.70	4.70
Blow room filter	0.60	0.60
Bale picking	0.03	0.03
Total	5.33	5.33
Cards dropping	2.60	Keep separate cards for comber section keeping total waste% 5.0–5.50
Flat strip	3.0	
Fan fly	0.40	
Cylinder fly	0.30	
Total	6.30	5.0–5.50
Sweeping	0.40	0.40
Hard waste	0.40	0.40
Noil	x	15.5–16.0
Grand total	12.43%	26.63–27.13%

8.3   Yarn costing

For calculating the costing of yarn, let us consider 20s combed cotton manufactured from J-34 R/G having a selling price of Rs. 200 per kg.

Selling rate – Rs. 200 per kg.

Selling expenses – Rs. 9.0 per kg.

So, Net realization – Rs. (200−9.0) = Rs. 191 per kg.

Row material cost = Rs. 126 per kg.

Recovery = 73%.

Clean material cost = Rs. 126/0.73 = Rs. 172.60.

Realization from total waste = Rs. 18.0 per kg.

Net clean raw material cost = Rs. (172.60 − 18) = Rs. 154.60.

Packing cost = Rs. 3.0 per kg.

Prime cost = (Net clean raw material cost + packing cost) = (154.60+3.0) = Rs. 157.60.

Contribution = Net realization − prime cost = Rs. (191–157.60) per kg = Rs. 33.40 per kg.

Ring frame production/480 spindle/day = 390 kg.

Contribution in terms of rupees = Rs. 33.40 × 390 = Rs. 13,026.

Fixed expenses per day per ring frame (480 spindles) = Rs. 10,300.

Fixed expenses/ring frame/day = wages + salary + power + depreciation + stores+ utilities + administrative cost, etc.

So, Profit/ring frame/day = Rs. (13,026 − 10,300) = Rs. 2726.

Profit/kg of yarn = Rs. 2726/390 = Rs. 6.989 or Rs. 7.0.

$$\text{Profit/spindle/shift} = \frac{390 \times 7}{480 \times 3} = \text{Rs. } 1.89.$$

Note: All data used in the above calculation are hypothetical.

8.3.1   Some basic ideas on how to fetch profit in yarn selling price in spinning mills

Following factors are to be considered seriously for achieving mill's profit.

1. Optimum grammage/spindle to be attained.
2. Spindle utilisation should be 99% and above.
3. Efficiency of the unit should be above 92%.
4. Yarn realisation for carded, combed and synthetic yarn should be 86–87%, 72–74% and above 97%, respectively.

5. Unusable waste% should be controlled.

6. Hard waste% should not go above 0.4%.

7. Proper utilisation of work force through optimum work load distribution is to be taken.

8. Wastage of power should be controlled and there should be overall effort to minimise power consumption.

    Besides above, there are other areas also but the prime responsibilities and accountabilities of a technical person should centre round the above points.

Main causes for low spindle utilisation are:

(i) power cut, (ii) worker shortage,(iii) frequent count change, (iv) more maintenance time taken than stipulated time, (v) back material shortage, (vi) poor supervision.

Spindle stoppage due to unavoidable reason:

(i) count change, (ii) electrical/mechanical maintenance, (iii) change pinion changes, (iv) programme changes, (v) R/Tr. changes, (vi) spindle tape replacement.

Spindle stoppage due to avoidable reasons:

(i) back material shortage, (ii) late start of machines, (iii) empties short, (iv) workers short.

8.4   Power saving

If we look back to 1980's decade, we observe while dissecting the cost structure of yarn that the raw material contribution was 50% on an average, salaries and wages were 8–9% and power cost was 7–8%. Other costs like selling, packing, administration, depreciation, etc. would come next in the race. But at present, power cost is taking the second position at the rate of 13–14%. Cost of raw material and salaries/wages have also increased but with a slower pace. Hence, saving of power by controlling the wastage has become very much imperative to make a spinning mill viable. For this purpose, each and every spinning mill should form a power saving cell constituting in charge of production, maintenance (Mechanical and Electrical), QA and local administrative head to see and implement the best way to control and save power consumption.

8.4.1   Steps to be taken for power saving

Power saving can be done through 3-tier approach, viz.:

1. Without spending any money but with a change of attitude and approach to save power.
2. By spending a small amount of money.
3. By spending a good amount of money with technological up gradation.

Above three approaches are discussed below:

8.4.1.1   Power saving without spending any money

Different steps to be taken for power saving is discussed below:

1st Point: At the very initial stage, we must try to know where we stand and which areas are consuming maximum power. Ring frames, humidification and ring doubling section are three major areas in this matter.

At first stage, we should check the motor load efficiency of each motor fixed to ringframe and ring doubling machine. A format as shown below is to be prepared (Table 8.5).

Table 8.5: Ring frame motor specifications

Department – Ring frame

Machine	M/C no	M/C H.P.	Motor kW	Actual kW	Motor load efficiency (%)
Ring frame	1	20	15	9.0	60

Motor load efficiency below 65% should be checked by electrical department to find out the reason for low motor load efficiency. It may happen that a particular motor needs rewinding.

2nd Point: Leakage of pneumafil inside ring frame department from any ring frame should be blocked completely. Hot air leakage from pneumafil box should never be allowed to persist. This hot air will change the RH in the department with humidity pocket and will also increase the dry bulb temperature. Consequently water from the air washer chamber has to be increased. This will increase more power consumption and simultaneously generate more wondering flies in the department with increased ends down.

3rd Point: Cotton spinning mills produce hosiery yarns besides producing weaving yarn with finer and coarser counts. We should make separate zones for ring frames manufacturing hosiery and coarser counts and for weaving and finer courts. Spindle speeds for hosiery and coarser courts are kept lower and as such we should use appropriate low H.P. Motor for these frames to achieve better motor load efficiency.

4th Point: Motor pulley and machine pulley alignment should be checked on all machines and required correction is to be done where found necessary.

V-belts should be present on all grooves of motor pulley and machine pulley with equal tension. Oversize belt or loose belt should be replaced by the new one and all belts should be equally tensioned.

5th Point: Height of tube lights should be brought down to a lower level for better lighting. If possible, remove one tube light alternatively from the overhead tube lights.

6th Point: If there is any break for tiffin in night shift, all pneumafil motors should be stopped during tiffin time.

7th Point: Tribological approach to save power(power consumption due to friction)

Mineral oil used as spindle oil generally tends to deposit impurities such as sediments, wear debris which give resistance for free movement of spindle and hence consumes more power. Dispersant oil gives no suspended materials accumulated at the foot step. Therefore, the frictional energy consumption is less.

During greasing of bottom fluted roller bearing, arbor, Jockey pulley, old greases are to be flushed out and then new grease to be forced into the place. Hard old greases give more resistance to free rotation of above parts and needs more frictional energy. Also, the amount of dispersant oil, which is to be given during spindle oil toppling into the bolster, be of measured fixed quantity. Excess or low will cause more power consumption due to friction effect.

A general consciousness is to be created among all staff and work force for the importance of power saving. Lights and fans are to be switched off while leaving the room of work place. A typical bad habit of workers and even with the supervisory staff to clean the clothes and body with the compressed air is to be stopped completely.

All the sprinkler heads for spraying water in air washer chamber should run. No damaged sprinkler should be used and there should be no chocking of any sprinkler. All dampers should work properly and air filters are to be cleaned daily.

In ring doubling machines, spindle speed should be kept low to such an extent that balancing with cheese winding is maintained. Excess capacity motor are to be replaced and proper care for belts alignment and number of v-belts, loose v-belts are to be taken care of like ring frame machines.

Spinning plan is to be prepared by taking full achievable standard efficiency of each machine starting from blow-room up to finishing stage and extra machines are to be kept out of production. Lower efficiency of any machines should never be tolerated.

8.4.1.2   *Power saving by spending a small amount of money*

Second tier approach consists of financial involvement with a small and affordable amount. These amounts mostly are to be spent for power auditing or saving processes. These are mainly:

(i)   Clip on power meter for measuring KW/KVA/PF, etc.

   Two types of instruments are available, viz.:

   (a) Instantaneous, (b) print out.

   Print out is required for blow room, ring frame and ring doubling.

(ii)   Thermometer (non-contact type)

   This is mainly required to measure the temperature of the motor which should be equal to ambient temperature +40 °C. If this temperature is higher than the above standard temperature, then the rewinding of armature is needed.

(iii)   Energy meter to conduct power audit in terms of watts/spindle for individual ring frame.

   Soliman [2] developed the following equations for power consumption in ring frame:

$$P_N \text{ (no load power)} = 3.33 \times g^{1.9} \times N^{1.4} \times h \times 10^{-4} \text{ watts/spindle.}$$

where, g = spindle gauge,

   h = lift,

   N = spindle speed in thousand.

$$P_S \text{ (spinning power)} = 4.25 \times G^{0.87} \times D_r \times N^{2.4} \times 10^{-5.}$$

$$P_N + P_S = \text{Power at the start of the doff.}$$

$$P_P \text{ (power with package)} = d^{3.5} \times h \times N^{3.1} \times 10^{-6}$$

where, d = package diameter in cm.

$$\text{Power at full doff} = P_N + P_S + P_P.$$

$$\text{Average power (full doff)} = P_S + 0.53P_P + P_N.$$

The following equation may be used for verification against energy meter reading[3]:

$$P = [7.5 + (80/c) + (500/t^2)]n^2 \text{ Watts/spindle,}$$

where, P = power consumption in watts/spindle.

   n = spindle speed, rpm/10,000,

   c = count in English (Ne),

   t = tpi.

The above equation is applicable to a package size of 7″ lift and 42 mm ring diameter. However, a package size dependent factor R is introduced in the above equation as:

$$P = [7.5 + (80/c) + (500/t^2)]n^2R \text{ watt/spindle.}$$

where, R = 1 for 7″ lift and 42 mm ring diameter.
                     = 0.90 for 7″ lift and 40 mm ring diameter.
                     = 0.80 for 6.5″ lift and 38 mm ring diameter.
                     = 0.70 for 6.0″ lift and 38 mm ring diameter.

8.4.1.3   Power saving by spending a good amount of money with technological up gradation

Third and last tier approach is based on the technological up gradation which involves good amount of financial involvement. Some of the suggested up gradation are:

(1) Good capacitor to improve the power factor.
(2) 18.5 mm/19.0 mm spindle wharve diameter.
(3) Low weight pulley.
(4) Cell type air washer.
(5) PVC eliminator.
(6) Fibre reinforced propeller for air, washer fan.
(7) Use of 36 Watts fluorescent tube lights.
(8) Use of electronic ballast for light fitting.
(9) Replacement of mercury vapour street light inside mill campus by sodium vapour light.
(10) Modernisation of machineries consuming less power with high productivity.

SITRA has developed a new regression equation[4], which gives better correlation between actual and estimated power consumption over Soliman's equation[2].

SITRA has measured the power at full doff stage in about 1500 ring frames and their machine andprocess variables (e.g. count, spindle weight, spindle wharves diameter, lift, ring diameter, speed) were analysed to form a multiple regression equation as follows.

$$P = -0.147C + 0.0146W_s + 0.275D_s - 0.003112L + 7.174D_R - 0.0785D_R^2 + 0.00345S_s - 0.0000000123S_s^2 - 1.655t + 0.0291t_2 - 167.$$

where,
$$P = \text{power in watts/spindle.}$$
$$C = \text{count in Ne.}$$
$$W_s = \text{weight of spindle ing.}$$
$$D_s = \text{wharves diameter of the spindle inmm.}$$
$$L = \text{lift in mm.}$$
$$D_R = \text{ring diameter in mm.}$$
$$S_s = \text{spindle speed in rpm.}$$
$$t = \text{twist per inch given in the yarn.}$$

8.4.2   Unit per kg (UKG) calculation

UKG is defined as the units required to produce 1 kg of yarn of 40sEnglishcount with a grammage of 88.8 per spindle per 8 h.

1. Blow room requires 325 units/8 h for 2 number of scutchers, each producing 1150 kg i.e.325 units required for 2300 kg of B/R production.
   Units required for 110 kg of B/R production = 325×110/2300 = 15.5.
   Note: To produce 100 kg of yarns, 110 kg of laps are required.

2. Cards require 23 units/shift for 110 kg production.
   From 110 kg of laps, we get 105 kg of carded sliver.
   For 110 kg of carding production, units required 23.
   Units required for 105 kg of carded silver = 23 × 105/110 = 22 units.

3. Drawing requires 60 units/shift for production of 300 kg × 2 deliveries production.
   104 kg of drawing sliver is obtained from 105 kg of carded slivers.
   So, units required for104 kg of production = 60 × 104/600 = 10.4 units.

4. Simplex requires 25 units for production of 216 kg/shift having 108 spindle.
   From 104 kg of drawing sliver, we get 103 kg of roving.
   Units required for 103 kg of roving = 25 × 103/216 = 11.90 units.

5. Ring frame requires 88 units/shift for a grammage of 88.8/spindle for 40s Ne.

   So, prod/ring frame (432 spindle) per 8 h in kg $= \dfrac{88.8 \times 432}{1000} = 38.36$ kg.

   Units required for 100 kg of yarns = 88 × 100/38.36 = 299.40 units
   Summary:

(a)	Blow room	–	15.50 units.
(b)	Carding	–	22.00 units.
(c)	Drawing	–	10.40 units.
(d)	Simplex	–	11.90 units.
(e)	Ring frame	–	229.40 units.
	Total	–	**289.20 units.**

   Units required for producing 1 kg of yarn (40s Ne) = 2.89.

6. Post spinning:
   • 50 kg will go for reeling.
   • 50 kg will go for cone winding.

(i) Reeling requires 2 unit/shift for a production of 30 kg/reeling machine/8 h.

Units required to produce 50 kg of reeled yarn = $2 \times 50/30 = 3.2$ units.

(ii) Cone winding which needs 36 units per shift for a production of 340 kg/shift having 120 drum. Units required for 50 kg $= \dfrac{50 \times 36}{340}$ = 5.30 units.

From reeling the processing sequences are (a) bundling and(b) baling.

(a) Bundling requires 100 units/shift for a production of 1400 kg/press.

Units required for 50 kg bundling = $100 \times 50/1400 = 0.36$ units.

(b) Baling requires 72 units/shift for a production of 3000 kg/shift.

Units required for 50 kg baling = $72 \times 50/3000 = 1.2$ units.

Units required for humidification for 100 kg of yarn = 40 units.

Units required for lighting for 100 kg of yarn = 15.

Total units required for 100 kg of yarn (40s Ne)

Blow room	–	15.50 units
Carding	–	22.00 units
Drawing	–	10.40 units
Simplex	–	11.90 units
Ring frame	–	229.40 units
Reeling	–	3.20 units
Bundling	–	0.36 units
Baling	–	1.20 units
Cone winding	–	5.30 units
Humidification	–	40.00 units
Lighting	–	15.00 units
Total	–	**354.26 units**

Units required producing one kg of 40s Ne yarn (UKG) = 3.54.

8.5   References

1. "Financial Management –Theory & Practice"by Prasanna Chandra, Tata McGraw Hill Education Private Limited,p. 898.

2. Soliman, H., Power requirements in cotton and worsted ring spinning, Ph.D. thesis, ETH, Zurich, 1963.

3. Prakasam, R., etal.,48th Joint Technological Conference, April2007, p. 37.

4. Prakasam, R., etal.,48th Joint Technological Conference, April2007, p. 36.

Pallet packing

Pallet packing is suitable for exports and these pallets are put in the containers which are loaded inside factory premises. This system of packing is cheaper than normal packing and there is least chance of damage of cones during transportation.

Pallet packing consists of a wooden platform having a dimension of $L \times W \times H = 44'' \times 44'' \times 6''$.

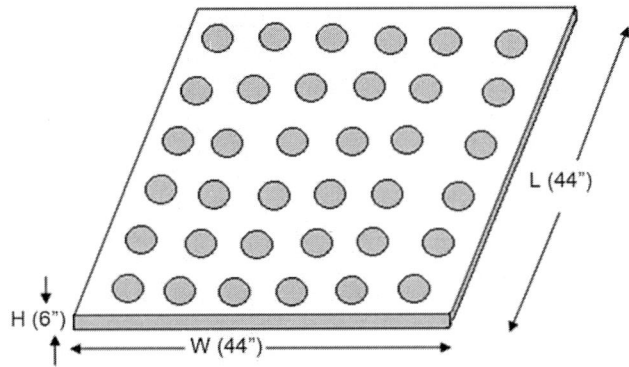

Figure A1: Wooden platform for pallet packing

where,

$$L = \text{length in inches,}$$
$$W = \text{width in inches,}$$
$$H = \text{height in inches.}$$

Total height of pallet after arranging cones in layers is $90'' + 6''$ (wooden platform) $= 96''$. 36 cones are arranged, as shown in figure in each layer and there are 14 such layers placed one above other.

So, total number of cones $= 36 \times 14 = 504$.

Cone size and specification:

(i) Empty paper cone size (taper) $= 3°30'$.

(ii) Cone length $= 165$ mm.

   (iii) Empty cone diameter = top – 23 mm, bottom –43 mm.
   (iv) Full cone diameter = top – 167 mm, bottom – 187 mm.
   (v) Empty paper cone weight = 32 g.
   (vi) Full cone weight = 2 kg.
   (vii) Weight of material on full cone = 1.97 kg.
   (viii) Weight of wooden platform = 38 kg.
   (ix) Total net weight = (504 × 1.97) = 992.9 kg.
   (x) Gross weight = (504 × 2) + 38 kg = 1046 kg.

Wooden platform is placed on the base plate of the Pallet Machine and 36 cones are arranged on the wooden platform in the manner as shown in Fig. A1. Next subsequent layers with 36 cones in each layer are arranged one above the other. In this way 14 layers are arranged on the wooden platform. The base plate of the Pallet Machine moves in a circular rotation and thus all the 14 layers of cones are wrapped by stretched cling film from the roll mounted on the carriage moving up and down, as shown in Fig. A2 above. These pallets are then put in a container (small container take 10 pallets andbig container takes 16 pallets) which is directly transported for shipment.

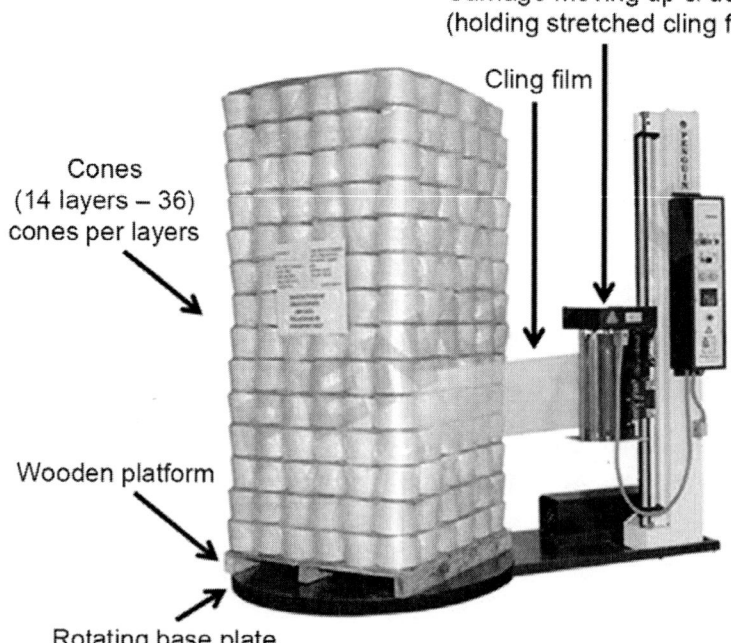

Figure A2: Pallet packing machines for cones

Few important suggestions for "material handling" in spinning mills

Materials handling is one of the most important practices in spinning mills that pays a direct dividend in yarn's quality in terms of thin places (frequent faults), yarns appearance, classimat faults, count cv%, etc. Materials handling doesnot confine to a single aspect i.e. "handling of materials", but it means how we:

- Transport the materials from one machine to other,
- Transport the materials from one department to other,
- Store the materials in the department.
- How the poor conditions of carding/drawing cans impair the fibres laying and arrangements and distribution of fibres in the slivers?
- How the poor work practices of the workers are damaging the fibres laying and distribution in fibre matrix?

In many mills, it was found the full simplex bobbins were kept on floor in a row one above another. Simplex bobbins were carried to ring frame department by placing simplex bobbins on stretched arms of workers being damaged by sweat of the worker's body. Excess ringnframe doffs are kept in a polyethylene bags as there is a shortage of doffing boxes, etc. Modernisation of spinning mills are taking care of this materials handling to a great extent through transport by link system.

Important relations:

$$\text{Twist factor} = \frac{\text{Twist per inch}}{\sqrt{N_e}}$$

$$\text{Twist multiplier} = \text{Twist per centimeter} \times \sqrt{\text{tex}}$$

$$\text{Cover factor} = \frac{\text{Threads/cm}}{\sqrt{\text{tex}} \times 10^{-2}}$$

Recommended SI units for textiles

Table 1

Sl. no.	Test parameter	SI units	Symbol
1	Length	Millimetre	mm
		Centimetre	cm
		Meter	m
2	Width	Millimetre	mm
		Centimetre	cm
3	Gauge length	Millimetre	mm

Contd...

Contd...

Sl. no.	Test parameter	SI units	Symbol
4	Thickness	Millimetre	mm
5	Linear density	Tex	tex
		Millitex	mtex
		Decitex	dtex
		Kilotex	ktex
6	Diameter	Micrometre	μm
		Millimetre	mm
7	Mass per unit area	Grams per square meter	g/m^2
8	Breaking load	Millinewton	mN
		Newton	N
9	Tenacity	Millinewton/tex	mN/tex
		Centinewton/tex	cN/tex
10	Specific stress	Centinewton/tex	cN/tex
11	Tearing strength	Newton	N
12	Bursting pressure	Kilonewton per square meter	kN/m^2
13	Bending rigidity	Millinewton	mN
		Square millimetre	mm^2

SI prefixes

The 20 SI prefixes used to form decimal multiples and submultiples of SI units are given in Table 2.

Name	Factor	Symbol	Name	Factor	Symbol
Yotta	10^{24}	Y	Deci	10^{-1}	d
Zetta	10^{21}	Z	Centi	10^{-2}	c
Exa	10^{18}	E	Milli	10^{-3}	m
Peta	10^{15}	P	Micro	10^{-6}	μ
Tera	10^{12}	T	Nano	10^{-9}	n
Giga	10^{9}	G	Pico	10^{-12}	p
Mega	10^{6}	M	Femto	10^{-15}	f
Kilo	10^{3}	k	Atto	10^{-18}	a
Hecto	10^{2}	h	Zepto	10^{-21}	z
Deka	10^{1}	da	Yocto	10^{-24}	y

Relation between SI units and conventional units

1 Newton = 102 g 1 cN = 1.02 g = 10 mN.

1 cN/tex=10 mN/tex = 1.02 g/tex = 0.113 g/demier.

Decitex × 0.9 = Denier or Denier × 1.11 = Decitex.

Relation between micronaire (µg/inch) and tex

$$\text{Tex} = \frac{\mu\text{g/inch}}{25.4}$$

or

$$\frac{\text{Denier}}{9} = \frac{\mu\text{g/inch}}{25.4}$$

or

$$\text{Denier} = \frac{9 \times (\mu\text{g/inch})}{25.4}$$

or

$$\text{Denier} = 0.354 \times (\mu\text{g/inch})$$

To find Denier from Micron:

$$\text{Diameter } (\mu) = 11.89 \sqrt{\frac{\text{Denier}}{\text{Density}}}$$

Above formula is needed particularly to convert fineness of wool in microns to Denier.

Deduction for establishing relation between:

(i) Newton and gram, and

(ii) cN/tex and g/denier.

We know $1b = 4.448$ N

or $1 \text{ N} = \dfrac{453.6 \text{ g}}{4.448} = 102 \text{ g}$

Again, we know,

$$\text{Tex} = \frac{\text{Denier}}{9}$$

or   1 g/denier $= 9$ g/tex   ...(1)

* (1 g = 0.98 CN) from relation (1) 1 g/denier = 9 × 0.98 CN/tex

= 8.82 CN/tex.

or $\text{CN/tex} = \dfrac{1}{8.82}$ g/denier

= 0.113 g/denier.